AF556200

Technology of Food Preservation

Technology of Food Preservation

Vikram Sharma

Technology of Food Preservation

ISBN 978-93-5111-520-5

Published in 2015 in India by

Reprint 2019

RANDOM PUBLICATIONS

4376-A/4B, Gali Murari Lal, Ansari Road
New Delhi-110 002
Phone : +9111-43580356, 011-23289044, 011-43142548
e-mail: sales@randompublications.com,
info@randompublications.com, randomexports@gmail.com

Type Setting by : Friends Media, Delhi-110089
Printed at : Mehra Printers, Delhi-110 092

Preface

Food preservation usually involves preventing the growth of bacteria, fungi and other micro-organisms as well as retarding the oxidation of fats which cause rancidity. It can also include processes which inhibit visual deterioration, such as the enzymatic browning reaction in apples after they are cut, which can occur during food preparation.

Many processes designed to preserve food will involve a number of food preservation methods. The processing of food is no longer simple or straightforward, but is now a highly inter-disciplinary science. A number of new techniques have developed to extend shelf-life, minimize risk, protect the environment, and improve functional, sensory, and nutritional properties. Preserving fruit by turning it into jam, for example, involves boiling, sugaring and sealing within an airtight jar. There are many traditional methods of preserving food that limit the energy inputs and reduce carbon footprint.

Maintaining or creating nutritional value, texture and flavour is an important aspect of food preservation, although, historically, some methods drastically altered the character of the food being preserved. In many cases these changes have now come to be seen as desirable qualities—cheese, yoghurt and pickled onions being common examples.

Containing fundamental and practical aspects of food preservation methods, the present book helps practising industrial and academic food scientists, technologists, and engineers develop high-quality, safe products through better understanding and control of the processes. It emphasizes practical, cost-effective, and safe-strategies for implementing preservation techniques and dissects the exact mode or mechanism involved in each method by highlighting the effects on food properties. It also presents a range of indirect approaches to improve quality and safety and good manufacturing practices.

Author

Contents

1

Introduction to Food Preservation

Preservation usually involves preventing the growth of bacteria, fungi (such as yeasts), and other micro-organisms (although some methods work by introducing benign bacteria, or fungi to the food), as well as retarding the oxidation of fats which cause rancidity. Food preservation can also include processes which inhibit visual deterioration, such as the enzymatic browning reaction in apples after they are cut, which can occur during food preparation.

Many processes designed to preserve food will involve a number of food preservation methods. Preserving fruit by turning it into jam, for example, involves boiling (to reduce the fruit's moisture content and to kill bacteria, yeasts, etc.), sugaring (to prevent their re-growth) and sealing within an airtight jar (to prevent recontamination). There are many traditional methods of preserving food that limit the energy inputs and reduce carbon footprint.

Maintaining or creating nutritional value, texture and flavour is an important aspect of food preservation, although, historically, some methods drastically altered the character of the food being preserved. In many cases these changes have now come to be seen as desirable qualities – cheese, yoghurt and pickled onions being common examples.

Since early historical time food preservation has been second only in importance to food production. Grapes and other fruits were dried by the ancients to preserve them; fruit juices were fermented to make wines and vinegars; cereals and vegetables were stored to protect them against moisture and decay; olives were preserved by salting; and meats were salted, dried, and smoked. The use of sugar and vinegar in preserving fruits and vegetables came later. The preservation of foods by sterilization in sealed containers is a development of the nineteenth century and dates from its discovery by Nicholas Appert in France about 1800. Cold storage, as a means of preserving all perishable products, has, during the past century, developed into a very great industry.

Three billion cans of food, valued retail at $600,000,000, were sold in the United States in 1916. The meat packing and cold storage industries compare favorably with the canning industries in size. The wholesale value of the raisin crop in California is over $10,000,000 annually. The other dried fruit industries are smaller but their aggregate value amounts to many millions of dollars yearly in the United States. From this, the importance of commercial food preservation may be seen.

Commercial food preservation cares for the bulk of the food products but beside the food so preserved, there are many millions of jars and cans of fruits and vegetables, glasses of jellies, jams, and marmalades and many thousands of hams and bacons "put up," by the housewife and farmer. Much food that would otherwise be wasted is saved and in addition a varied diet throughout the year at low cost is made available in many homes.

Why Food Spoils?

Food spoils because of the growth and destructive action of microscopic living organisms. They are cornmonly termed “germs.” The various methods of food preservation are practically all based upon processes that destroy these organisms or prevent their growth and activity. Because they are microscopic and because they are living organisms, we shall for convenience call them “microorganisms.”

Molds

The molding of food is a common phenomenon. In some cases the food is completely spoiled; in others, the decomposition is not sufficient to make the product inedible; and in a few products, the growth of certain molds is desirable.

The most prevalent mold and the one causing the most damage is the “blue mold,” otherwise known as “Penicillium expansum.” It first appears usually as a white cottony growth on fruits, cheese, cured meats, vegetables, jellies, wine tanks, leather left in dark closets, and on other articles favorable to its growth. This cottony growth of mold threads is known as a "mycelium." Later, the mold becomes "powdery" in appearance and green or blue in color. This change in color is due to the formation of billions of microscopic cells or "spores." The spores are very light and easily detached. They are carried by the air or other agencies from place to place. They are floating in the air at all times and places and are present on the surfaces of all fresh foods. They are capable of sprouting when conditions become favorable. A large growth may start from a single cell or spore.

Fruits whose skins become broken in transit suffer badly from this mold and acquire a moldy taste and odor. In some such cases the growth will not be apparent because the mold threads are growing in the pulp or juice of the fruit.

The surface of jellies may become overgrown by this organism and the upper portion of the jelly completely spoiled. Leaky jars of fruit may mold from the growth of penicillium spores gaining entrance through the leaks. Bacon and cheese may develop green spots of this mold on the surface and still not be spoiled if the mold is removed in time. The inside of wine or vinegar barrels may be completely spoiled where this mold is allowed to develop through improper care of the barrels.

The blue mold can be controlled, but great care must be taken if it is to be completely eliminated. Its spores are killed by heating to 180° F. and. growth is prevented by many chemicals.

"Black Mold," otherwise known as "Aspergillus niger," often occurs on fruits that have become moist on the surface or broken; or it may occur on other products occasionally. It does not produce a moldy taste or odor; it is much less prevalent, and is easier to control than is the blue mold.

"Pin Mold," or "Gray Mold," or "Bread Mold," usually causes the molding of bread stored in a moist place. It also occurs frequently on fruits and may appear as "whiskers" on peaches, grapes, and other fruits, shipped long distances in boxes. It is not especially important in food preservation. It is known botanically as Mucor.

There are hundreds of other forms of molds but the above forms are by far the most common on food products.

Molds are not always deleterious in their action. Camembert, Rocquefort, and some other fancy cheeses owe their distinctive quality to the growth of special forms of Penicilliurn molds. A form of Aspergillus mold, known as Aspergillus oryzae, is used extensively in Japan in making "Saki," Japanese beer. A Mucor mold is used frequently in distilleries in the production of alcohol from cereals.

In general, molds are of interest in food preservation because of their capacity for spoiling food, their universal occurrence on food products, and the difficulty in killing their spores by heat or controlling their growth in other ways.

Yeasts

When a fruit juice is allowed to stand a few days it undergoes fermentation. The sugar is destroyed and alcohol and carbonic acid gas are formed. This change is brought about by another group of microscopic organisms, known as yeasts. Yeasts are used in bread making, vinegar manufacture, and in the production of various fermented beverages.

Unlike molds, they do not form a mycelium, i. e., a thread-like growth, but only develop as microscopic cells of various forms. They appear in fermented liquids as a white sediment or a cloudy growth throughout the liquid.

They are universally present in the air, on the surfaces of fruits, vegetables, and of tables, knives, etc., and are capable of growing in and spoiling sugary liquids, crushed fruits, jellies that do not have sufficient sugar, and in other products containing from one to 65% sugar. More sugar than 65% prevents their growth.

Jars and cans of fruit that become leaky after sterilization become infected with yeast cells carried in by air passing into the containers. Growth and fermentation take place and the pressure of. the carbonic acid formed by the yeast causes the container to swell or burst. Much canned fruit is lost in this way. The housewife usually attributes the loss to the entrance of air. It is in reality caused by yeast gaining entrance with the air; air alone would be incapable of causing fermentation.

Yeasts are easily killed by heat, a temperature of 60° C. or 140° F. being sufficient, and in general, yeasts are more easily controlled than molds. Conditions that will eliminate molds will also remove yeasts.

Yeasts cause the "souring," "working," or fermenting of spoiled jars or cans of fruit, bottles of fruit juices, or glasses of jelly. They are therefore of much importance in the preservation of fruit products.

They are necessary in the manufacture of all fermented beverages, denatured alcohol, vinegar, and yeast-risen bread. Yeasts are the most useful of all the microorganisms met with in food preservation.

Bacteria

Milk sours on standing; meat and many cooked vegetables putrefy unless spoiling is prevented; dill pickles and sauerkraut undergo certain characteristic changes. These changes are wholly, or in most part, brought about by bacteria. They comprise the third main group of "germs" or microorganisms. Like the other two groups they are universally distributed. Bacteria are, as a rule, smaller than yeasts and differ from them in their method of reproduction. Yeasts reproduce by budding and bacteria by splitting in two, i. e., by "fission." Bacteria prefer nitrogenous substances of low acid content, such as milk, meat, peas, and beans, and do not grow readily on fruits or acid vegetables. Molds and yeasts prefer sugary, acid materials.

Yeast and mold spores are easily killed by temperatures below 212° F. Many bacterial spores survive temperatures above 212° F., the boiling point of water. For this reason, many foods containing such spores are exceedingly difficult to sterilize by heat, This does not apply to foods high in acid because these bacteria can not grow readily in the presence of much acid and are more easily killed in acid foods.

Yeasts and molds produce relatively harmless compounds in food products. Bacteria on the other hand may produce in canned vegetables, in meats, and in cheese, extremely poisonous compounds. These are the ptomaines and botulinus poison. It is therefore necessary to be sure that such products as canned peas, beans, corn, and meats, are thoroughly sterilized, in order that poisoning will not occur.

Several forms of bacteria are extremely useful in food preservation and food manufacture. The two most important are vinegar bacteria, necessary in making vinegar, and lactic acid bacteria, essential in the manufacture and preservation of sauerkraut, pickled green olives, silage, and cheese. " Vinegar Mother " is a growth of vinegar bacteria; the sour taste of sauerkraut and sour milk is brought about by lactic acid formed by lactic acid bacteria.

Spoiling of Foods by Chemical and Physical Changes

Some food products decompose without the action of organisms. Edible fats and oils become rancid through the action of the oxygen of the air. Meats are sometimes practically spoiled by the use of too much salt in salt curing. Dried fruits may be greatly injured by leaving them too long in the sun on trays. Canned goods sometimes act upon the tin of the cans to such an extent that they become poisonous or inedible.

Practically all food products undergo slow changes through drying or oxidation when left exposed to the air. Even cereals deteriorate with age in bins, elevators, etc.

Changes of this sort are as a rule slower and more easily controlled than bacterial changes. It is usually only necessary to exclude moisture or air or control the temperature.

Methods of Food Preservation

Drying

Drying is one of the most ancient food preservation techniques, which reduces water activity sufficiently to prevent bacterial growth.

Refrigeration

Refrigeration preserves food by slowing down the growth and reproduction of micro-organisms and the action of enzymes which cause food to rot. The introduction of commercial and domestic refrigerators drastically improved the diets of many in the Western world by allowing foods such as fresh fruit, salads and dairy products to be stored safely for longer periods, particularly during warm weather.

Freezing

Freezing is also one of the most commonly used processes commercially and domestically for preserving a very wide range of food including prepared food stuffs which would not have required freezing in their unprepared state. For example, potato waffles are stored in the freezer, but potatoes themselves require only a cool dark place to ensure many months' storage. Cold stores provide large volume, long-term storage for strategic food stocks held in case of national emergency in many countries.

Vacuum packing

Vacuum-packing stores food in a vacuum environment, usually in an air-tight bag or bottle. The vacuum environment strips bacteria of oxygen needed for survival, slowing spoiling. Vacuum-packing is commonly used for storing nuts to reduce loss of flavor from oxidation.

Salt

Salting or curing draws moisture from the meat through a process of osmosis. Meat is cured with salt or sugar, or a combination of the two. Nitrates and nitrites are also often used to cure meat and contribute the characteristic pink color, as well as inhibition of Clostridium botulinum.

Sugar

Sugar is used to preserve fruits, either in syrup with fruit such as apples, pears, peaches, apricots, plums or in crystallized form where the preserved material is cooked in sugar to the point of crystallisation and the resultant product is then stored dry. This method is used for the skins of citrus fruit (candied peel), angelica and ginger.

Smoking

Smoking is used to lengthen the shelf life of perishable food items. This effect is achieved by exposing the food to smoke from burning plant materials such as wood. Most commonly subjected to this method of food preservation are meats and fish that have undergone curing. Fruits and vegetables like paprika, cheeses, spices, and ingredients for making drinks such as malt and tea leaves are also smoked, but mainly for cooking or flavoring them. It is one of the oldest food preservation methods, which probably arose after the development of cooking with fire.

Artificial food additives

Preservative food additives can be antimicrobial; which inhibit the growth of bacteria or fungi, including mold, or antioxidant; such as oxygen absorbers, which inhibit the oxidation of food constituents. Common antimicrobial preservatives include calcium propionate, sodium nitrate, sodium nitrite, sulfites (sulfur dioxide, sodium bisulfite, potassium hydrogen sulfite, etc.) and disodium EDTA. Antioxidants include BHA and BHT. Other preservatives include formaldehyde (usually in solution), glutaraldehyde (kills insects), ethanol and methylchloroisothiazolinone.

Pickling

Pickling is a method of preserving food in an edible anti-microbial liquid.

Pickling can be broadly categorized into two categories: chemical pickling and fermentation pickling.

In chemical pickling, the food is placed in an edible liquid that inhibits or kills bacteria and other micro-organisms. Typical pickling agents include brine (high in salt), vinegar, alcohol, and vegetable oil, especially olive oil but also many other oils. Many chemical pickling processes also involve heating or boiling so that the food being preserved becomes saturated with the pickling agent. Common chemically pickled foods include cucumbers, peppers, corned beef, herring, and eggs, as well as mixed vegetables such as piccalilli.

In fermentation pickling, the food itself produces the preservation agent, typically by a process that produces lactic acid. Fermented pickles include sauerkraut, nukazuke, kimchi, surströmming, and curtido. Some pickled cucumbers are also fermented.

Lye

Sodium hydroxide (lye) makes food too alkaline for bacterial growth. Lye will saponify fats in the food, which will change its flavor and texture. Lutefisk uses lye in its preparation, as do some olive recipes. Modern recipes for century eggs also call for lye.

Canning and bottling

Canning involves cooking food, sealing it in sterile cans or jars, and boiling the containers to kill or weaken any remaining bacteria as a form of sterilization. It was invented by Nicolas Appert. Foods have varying degrees of natural protection against spoilage and may require that the final step occur in a pressure cooker. High-acid fruits like strawberries require no preservatives to can and only a short boiling cycle, whereas marginal fruits such as tomatoes require longer boiling and addition of other acidic elements. Low acid foods, such as vegetables and meats require pressure canning. Food preserved by canning or bottling is at immediate risk of spoilage once the can or bottle has been opened.

Lack of quality control in the canning process may allow ingress of water or micro-organisms. Most such failures are rapidly detected as decomposition within the can causes gas production and the can will swell or burst. However, there have been examples of poor manufacture (underprocessing) and poor hygiene allowing contamination of canned food by the obligate anaerobe Clostridium botulinum, which produces an acute toxin within the food, leading to severe illness or death. This organism produces no gas or obvious taste and remains undetected by taste or smell. Its toxin is denatured by cooking, though. Cooked mushrooms, handled poorly and then canned, can support the growth of Staphylococcus aureus, which produces a toxin that is not destroyed by canning or subsequent reheating.

Jellying

Food may be preserved by cooking in a material that solidifies to form a gel. Such materials include gelatine, agar, maize flour and arrowroot flour. Some foods naturally form a protein gel when cooked such as eels and elvers, and sipunculid worms which are a delicacy in Xiamen in Fujian province of the People's Republic of China. Jellied eels are a delicacy in the East End of London where they are eaten with mashed potatoes. Potted meats in aspic, (a gel made from gelatine and clarified meat broth) were a common way of serving meat off-cuts in the UK until the 1950s. Many jugged meats are also jellied.

Jugging

Meat can be preserved by jugging, the process of stewing the meat (commonly game or fish) in a covered earthenware jug or casserole. The animal to be jugged is usually cut into pieces, placed into a tightly-sealed jug with brine or gravy, and stewed. Red wine and/or the animal's own blood is sometimes added to the cooking liquid. Jugging was a popular method of preserving meat up until the middle of the 20th century.

Irradiation

Irradiation of food is the exposure of food to ionizing radiation; either high-energy electrons or X-rays from accelerators, or by gamma rays (emitted from radioactive sources as Cobalt-60 or Caesium-137). The treatment has a range of effects, including killing bacteria, molds and insect pests, reducing the ripening and spoiling of fruits, and at higher doses inducing sterility. The technology may be compared to pasteurization; it is sometimes called 'cold pasteurization', as the product is not heated. Irradiation is not effective against viruses or prions, it cannot eliminate toxins already formed by microorganisms.

The radiation process is unrelated to nuclear energy, but it may use the radiation emitted from radioactive nuclides produced in nuclear reactors. Ionizing radiation is hazardous to life (hence its usefulness in sterilisation); for this reason irradiation facilities have a heavily shielded irradiation room where the process takes place. Radiation safety procedures ensure that neither the workers in such facility nor the environment receive any radiation dose from the facility. Irradiated food does not become radioactive, and national and international expert bodies have declared food irradiation as wholesome. However, the wholesomeness of consuming such food is disputed by opponents and consumer organizations. National and international expert bodies have declared food irradiation as 'wholesome'; UN-organizations as WHO and FAO are endorsing to use food irradiation. International legislation on whether food may be irradiated or not varies worldwide from no regulation to full banning. Irradiation may allow lower quality or contaminated foodstuffs to be rendered marketable.

It is estimated that about 500,000 tons of food items are irradiated per year worldwide in over 40 countries. These are mainly spices and condiments with an increasing segment of fresh fruit irradiated for fruit fly quarantine. food preservation is very good process

Pulsed electric field electroporation

Pulsed electric field (PEF) electroporation is a method for processing cells by means of brief pulses of a strong electric field. PEF holds potential as a type of low temperature alternative pasteurization process for sterilizing food products. In PEF processing, a substance is placed between two electrodes, then the pulsed electric field is applied. The electric field enlarges the pores of the cell membranes which kills the cells and releases their contents. PEF for food processing is a developing technology still being researched. There have been limited industrial applications of PEF processing for the pasteurization of fruit juices.

Modified atmosphere

Modifying atmosphere is a way to preserve food by operating on the atmosphere around it. Salad crops which are notoriously difficult to preserve are now being packaged in sealed bags with an atmosphere modified to reduce the oxygen (O_2) concentration and increase the carbon dioxide (CO_2) concentration. There is concern that although salad vegetables retain their appearance and texture in such conditions, this method of preservation may not retain nutrients, especially vitamins. Grains may be preserved using carbon dioxide by one of two methods; either using a block of dry ice placed in the bottom and the can is filled with grain or the container can be purged from the bottom by gaseous carbon dioxide from a cylinder or bulk supply vessel.

Carbon dioxide prevents insects, and depending on concentration, mold, and oxidation from damaging the grain. Grain stored in this way can remain edible for five years.

Nitrogen gas (N_2) at concentrations of 98% or higher is also used effectively to kill insects in grain through hypoxia. However, carbon dioxide has an advantage in this respect as it kills organisms through hypercarbia and depending on concentration hypoxia and, requiring concentrations of above 35%, or so. This makes carbon dioxide preferable for fumigation in situations where a hermetic seal cannot be maintained.

Air-tight storage of grains (sometimes called hermetic storage) relies on the respiration of grain, insects and fungi which can modify the enclosed atmosphere sufficiently to control insect pests. This is a method of great antiquity, as well as having modern equivalents. The success of the method relies on have the correct mix of sealing, grain moisture and temperature.

Nonthermal plasma

This process subjects the surface of food to a 'flame' of ionised gas molecules such as helium or nitrogen. This causes micro-organisms to die off on the surface.

High pressure food preservation

High pressure food preservation or pascalization refers to the use of a food preservation technique which makes use of high pressure. "Pressed inside a vessel exerting 70,000 pounds per square inch (480 MPa) or more, food can be processed so that it retains its fresh appearance, flavour, texture and nutrients while disabling harmful microorganisms and slowing spoilage. By 2005 the process was being used for products ranging from orange juice to guacamole to deli meats and widely sold.

Burial in the ground

Burial of food can preserve it due to a variety of factors: lack of light, lack of oxygen, cool temperatures, pH level, or desiccants in the soil. Burial may be combined with other methods such as salting or fermentation. Most foods can be preserved in soil that is very dry and salty (thus a desiccant), or soil that is frozen.

Many root vegetables are very resistant to spoilage and require no other preservation than storage in cool dark conditions, for example by burial in the ground, such as in a storage clamp. Century eggs are created by placing eggs in alkaline mud (or other alkaline substance) resulting in their "inorganic" fermentation through raised pH instead of spoiling. The fermentation preserves them and breaks down some of the complex, less flavorful proteins and fats into simpler more flavorful ones. Cabbage was traditionally buried in the fall in northern farms in the USA for preservation. Some methods keep it crispy while other methods produce sauerkraut[citation needed]. A similar process is used in the traditional production of kimchi. Sometimes meat is buried under conditions which cause preservation. If buried on hot coals or ashes, the heat can kill pathogens, the dry ash can desiccate, and the earth can block oxygen and further contamination. If buried where the earth is very cold, the earth acts like a refrigerator.

Controlled use of micro-organism

Some foods, such as many cheeses, wines, and beers will keep for a long time because their production uses specific micro-organisms that combat spoilage from other less benign organisms. These micro-organisms keep pathogens in check by creating an environment toxic for themselves and other micro-organisms by producing acid or alcohol. Starter micro-organisms, salt, hops, controlled (usually cool) temperatures, controlled (usually low) levels of oxygen and/or other methods are used to create the specific controlled conditions that will support the desirable organisms that produce food fit for human consumption.

Biopreservation

Biopreservation is the use of natural or controlled microbiota or antimicrobials as a way of preserving food and extending its shelf life. Beneficial bacteria or the fermentation products produced by these bacteria are used in biopreservation to control spoilage and render pathogens inactive in food. It is a benign ecological approach which is gaining increasing attention.

Of special interest are lactic acid bacteria (LAB). Lactic acid bacteria have antagonistic properties which make them particularly useful as biopreservatives. When LABs compete for nutrients, their metabolites often include active antimicrobials such as lactic and acetic acid, hydrogen peroxide, and peptide bacteriocins. Some LABs produce the antimicrobial nisin which is a particularly effective preservative.

These days LAB bacteriocins are used as an integral part of hurdle technology. Using them in combination with other preservative techniques can effectively control spoilage bacteria and other pathogens, and can inhibiting the activities of a wide spectrum of organisms, including inherently resistant Gram-negative bacteria.

Hurdle technology

Hurdle technology is a method of ensuring that pathogens in food products can be eliminated or controlled by combining more than one approach. These approaches can be thought of as "hurdles" the pathogen has to overcome if it is to remain active in the food. The right combination of hurdles can ensure all pathogens are eliminated or rendered harmless in the final product.

Hurdle technology has been defined by Leistner as an intelligent combination of hurdles which secures the microbial safety and stability as well as the organoleptic and nutritional quality and the economic viability of food products. The organoleptic quality of the food refers to its sensory properties, that is its look, taste, smell and texture.

Examples of hurdles in a food system are high temperature during processing, low temperature during storage, increasing the acidity, lowering the water activity or redox potential, or the presence of preservatives or biopreservatives. According to the type of pathogens and how risky they are, the intensity of the hurdles can be adjusted individually to meet consumer preferences in an economical way, without sacrificing the safety of the product.

Importance of Food Preservation

Food production and supply does not always tally with the demand or meets of the people. In some places there is surplus production of a food product, whereas in some other place there is inadequate supply. Even foods are perishable and semi-perishable like juicy fruits, vegetables, mangoes, tomato, papaya and many more, which very quickly gets spoilt. It is therefore important to improve and expand facilities for storage and preservation of food. Food preservation helps in:

1. Increasing the self-life of foods thus increasing the supply. So many perishable foods can be preserved for a long time.
2. Making the seasonal food available throughout the year.
3. Adding variety to the diet.
4. Saving time by reducing preparation time and energy, as the food has already been partially processed.
5. Stabilising prices of food, as there is less scope of shortage of supply to demand.
6. Decreasing wastage of food by preventing decay or spoilage of food.
7. Improving the nutrition of the population. Preserved foods help people to bring a variety in the diet, thereby decreasing nutritional inadequacies.

Development and Future Food Preservation Technology

Some of the future methods of food preservation are irradiation and chemical additives. Although these methods are currently in use, they are expected to expand and develop further. Irradiation of food is the process of exposing food to ionizing radiation. This process can alter the bacteria, microorganism or virus' DNA, without harming the food. Irradiation is attractive because of its selective targeting. It is already used on non-food items. "The molecular bonds in the microbial DNA are the main targets of the irradiation, but DNA and RNA synthesis, denaturation of enzymes and cell membrane alterations may also be affected". The process of irradiation opens the possibility to process a large number of foods in great quantities, however it can be expensive. The buildings for such a process require specific infrastructure and construction that is both expensive and time consuming. "The Food and Agriculture Organization (FAO), the International Atomic Energy Agency (IAEA), and World Health Organization (WHO) concluded in their report, that any food irradiated up to a maximum dose of 10 kGy is considered safe and wholesome". Essentially three things were concluded in their report: 1. It won't lead to toxicological changes in the food that will negatively affect our health, 2. The technology won't increase the microbial risk of the consumer, and 3. Irradiation won't lead to nutritional losses. However, most consumers are still radiation-phobic despite these positive results. They aren't willing to buy something that has, in their mind, been radiated and burned to safety. Another process is the addition of chemicals to preserve the freshness of the food. Consumers feel that this is a risk because of the unnatural state the food and their bodies are subjected to. New research has shown that even the most susceptible foods such as a ham and pepperoni sandwiches can be made to last years.

Lauren Oleksyk, leader of the food-processing, engineering and technology team at the U.S. Department of Defense Combat Feeding Directorate mixed water-absorbing ingredients including glycerol and sorbitol into the filling. They also increased the use of fine, edible polymer films, which are undetectable in the mouth. These were added to combat moisture that would be detrimental to the sandwich as it provides a good breeding ground for microorganisms and bacteria. In addition to the water absorbing ingredients, Oleksky's team also added oxygen eliminators. It sounds more intimidating than it is; essentially it is the process of adding oxygen-scavenging chemicals. Oxygen increases the possibility of spoilage to the food. By adding these chemicals, it limits the microorganism's chance to grow and thrive, and also inhibits the sandwich's ability to change chemically.

A potentially groundbreaking and safe process of food preservation is high pressure preservation. Dr. V.M. Balasubramaniam, a food engineer from Ohio State University, is developing a process where food is attacked with 100,000 pounds per square inch of pressure.

The second compartment is filled with hydraulic fluid, which is used to press the piston between the two compartments. When even more hydraulic fluid is pushed into the second compartment, the piston is pushed up into the first compartment. The water becomes more condensed, and as a result the pressure in the first compartment increases rapidly. After only a few minutes, any food can be ready to eat. While every other process we have discussed thus far has had some negative side effects, HPP appears to be the perfect solution. Because no chemicals are added, there is no contamination or taste alteration to the food. The bacteria in the food remain intact; however they die due to the pressure-induced dismantlement of their DNA structure. Almost all foods are able to be pressurized in the machine. The only exception is some veggies and fruits, which get too pulverized. The only major drawback to HPP is simple – the cost. High-pressure machines typically cost $3 million. This technology is so new, and therefore companies have not found a cheap and efficient way to mass produce these machines. However, with due time, we expect to see high pressure food preservation machines available to the public at a reasonable cost. This technology is simply too good to reserve for large industrial food processing plants.

In conclusion, food preservation has been pivotal to our society since the beginning. Preservation has come from simple processes such as salting, to more complex preserving methods such as irradiation and chemical additives. Looking into the future, High Pressure Preservation seems to be the next logical step. In an increasingly health-conscious world, an uncontaminated, well preserved food source is the ultimate goal, and hopefully the costs of production will decrease for all of us to enjoy HPP foods.

References

Alasalvar C. 2010. *Seafood Quality, Safety and Health Applications* John Wiley and Sons.

Alzamora SM, Tapia MS and López-Malo A. 2000. *Minimally processed fruits and vegetables: fundamental aspects and applications* Springer.

FAO: *Preservation techniques Fisheries and aquaculture department,* Rome. Updated 27 May 2005. Retrieved 14 March 2011.

Lee S. 2004. "Microbial Safety of Pickled Fruits and Vegetables and Hurdle Technology" *Internet Journal of Food Safety*, 4: 21-32.

Leistner I. 2000. "Basic aspects of food preservation by hurdle technology" *International Journal of Food Microbiology*, 55:181–186.

Yousef AE and Carolyn Carlstrom C. 2003. *Food microbiology: a laboratory manual* Wiley.

2

Preservation of Fruits and Vegetables

The increasing popularity of minimally processed fruits and vegetables has resulted in greater health benefits. Furthermore, the ongoing trend has been to eat out and to consume ready-to-eat foods. With this increasing demand for ready-to-eat, fresh, minimally processed foods, including processed fruits and vegetables preserved by relatively mild techniques, new ecology routes for microbial growth have emerged. In order to minimize the loss of quality and to control microbial growth, and thus ensure product safety and convenience, a hurdle approach appears to be the best method. According to Alzamora et al., hurdle technology can be applied several ways in the design of preservation systems for minimally processed foods at various stages of the food chain:

— As a "backup" measure for existing minimally processed products with short shelf life, to diminish microbial pathogenic risk and/or increase shelf life (i.e., use of natural antimicrobials or other stress factors, in addition to refrigeration).

— As an important tool for improving the quality of long shelf life products without diminishing their microbial stability/safety (i.e., use of heat coadjuvants to reduce the severity of thermal treatments).

— As a synergist. According to Leistner, in food preserved by hurdle technology, the possibility exists that different hurdles in a food will not just have an additive effect on stability, but could act synergistically. A synergist effect could work if the hurdle in a food hits different targets (e.g., cell membrane, DNA, enzyme systems, pH, a_w, Eh) within the microbial cell, and thus disturbs the homeostasis of the microorganisms present in several aspects. Therefore, employing different hurdles in the preservation of a particular food should be an advantage, because microbial stability could be achieved with a combination of gentle hurdles. In practical terms, this could mean that it is more effective to use different preservatives in small amounts in a food than only one preservative in large amounts, because different preservatives might hit different targets within the bacterial cell, and thus act synergistically.

During the last decade, minimally processed high moisture fruit products (HMFP), which are ambient stable (with $a_w > 0.93$), have been developed in seven Latin American countries, under the leadership of Argentina, Mexico, and Venezuela. This novel technology was successfully applied to peach halves, pineapple slices, mango slices and purée, papaya slices, chicozapote slices, banana purée, plum, passion fruit, tamarind, whole figs, strawberries, and pomalaca. The methodology employed was based on combinations of mild heat treatments, such as blanching for 1-3 minutes with saturated steam, slightly reducing the a_W (0.98-0.93) by addition of glucose or sucrose, lowering the pH (4.1-3.0) by addition of citric or phosphoric acid, and adding antimicrobials (1000 ppm of potassium sorbate or sodium benzoate, as well as 150 ppm of sodium sulphite or sodium bisulphite) to the product syrup. During storage of HMFP, the sorbate and sulphite levels decreased, as well as a_w levels, due to hydrolysis of glucose.

Trade of Fruits and Vegetables: Trends of Trade Consumption

Recently, the Food Agricultural Organization of the United Nations (FAO) predicted that the world population would top eight billion by the year 2030. Therefore, the demand for food would increase dramatically. As stated in the FAO report, "Agriculture: Towards 2015/30", remarkable progress has been made over the last three decades towards feeding the world. While global population has increased over 70 percent, per capita food consumption has been almost 20 percent higher. In developing countries, despite a doubling of population, the proportion of those living in chronic states of under nourishment was cut in half, falling to 18 percent in 1995/97. According to the report, crop output is projected to be 70 percent higher in 2030 than current output. Fruits and vegetables will play an important role in providing essential vitamins, minerals, and dietary fibre to the world, feeding populations in both developed and developing countries.

In developed countries, the U.S. continues to dominate the international trade of fruits and vegetables, and is ranked number one as both importer and exporter, accounting for approximately 18 percent of the $40 billion (USD) in fresh produce world trade. As a group, the European Union (EU) constitutes the largest player, with 15 additional export and import commodities contributing about 20 percent to total fresh fruit and vegetable trade. Within Europe, Germany is the principal exporter; Spain is the principal supplier; and the Netherlands plays an important role in the physical distribution process. In the Southern Hemisphere, Chile, South Africa, and New Zealand have become major suppliers in the international trade of fresh fruit commodities, although they remain insignificant in vegetable trade.

FAO estimated that the world production of fruits and vegetables over a three-year period (1993-1995) was 489 million tons for vegetables and 448 million tons for fruits. This trend increased as expected, reaching a global production of 508 million tons for vegetables and 469 tons for fruits in 1996. This trend in production is expected to increase at a rate of 3.2 percent per year for vegetables and 1.6 percent per year for fruits. However, this trend is not uniform worldwide, especially in developing countries where the lack of adequate infrastructure and technology constitutes the major drawback to competing with industrialized countries. Nevertheless, developing countries will continue to be the leaders in providing fresh exotic fruits and vegetables to developed countries. Most developing countries have experienced a high increase in fruit and vegetable production, as in the case of Asia (China) and South America (Brazil, Chile). Asia is the leading producer of vegetables with a 61 percent total volume output and a yearly growth of 51 percent. However, the U.S. continues to lead in the export of fresh fruits and vegetables worldwide with orange, grapes, and tomatoes. Brazil dominates the international trade of frozen orange juice concentrate, while Chile has become the major fresh fruit exporter with a production volume of 45 percent. Despite the large growth in exports in the 1990s, the U.S. remains a net importer of horticultural products. As U.S. consumers have become more willing to try new fruit and vegetable varieties, the imported share of the domestic market has increased. According to a USDA report, the total value of horticultural products imported into the U.S. has grown by more than 50 percent since 1990. If long-term projections hold for the next decade, the U.S. could achieve a trade balance surplus in horticultural products, due mainly to a global increase in the market. While the import value of horticultural products is projected to grow at a steady rate of 4 percent per year, between 1998 and 2007, the USDA's baseline projection period for exports are projected to grow by 5 to 7 percent per year.

The top six fruit producers, in declining order of importance, are China, India, Brazil, USA, Italy, and Mexico. China, India, and Brazil account for almost 30 percent of the world's fruit supply, but since most of this production is destined for domestic consumption its impact on world trade is minimal.

Traditional Consumption

Fruit and vegetable consumption per capita showed an increase of 0.38 percent for fresh fruits and 0.92 percent for vegetables per capita from 1986 to 1995. The highest consumption of fresh fruits was registered in China (6.4%), as the apparent per capita consumption of vegetables in China went from 68.7 kg per capita in 1986 to 146 kg in 1995 (53.8% growth rate), while African and Near East Asian countries showed a decrease in fresh fruit consumption. The lowest consumption of vegetables per capita was registered in Sub-Saharan Africa (29 kg of vegetables consumed both in 1986 and 1995). According to trade sources, Chinese customers purchased most of their fresh fruit at street retail shops and market places where imported fresh fruits are available and U.S. and European brand names have received recognition. Products such as Red Delicious apples, Sunkist oranges, and Red Globe table grapes are especially popular. Sunkist is one of the few brands of oranges consumers recognize. The trend toward fresh vegetable consumption in developing countries is one indication of the population's standard of living, but generally, fresh vegetables lose their market share to processed products. Many vegetables can be processed into canned products that cater to local tastes, (e.g., cucumbers and peppers). Easy to carry and convenient to serve, they can be stored for a long time, reducing losses incurred from the seasonal supply of surplus vegetables marketed yearly at the same time. Urban population is exploding in developing countries, having risen from 35 percent of the total population in 1990, and projected to rise 54 percent in 2020. With increasing urban populations, more free markets and wholesale markets will be required to increase the supply of fresh fruits and vegetables. For example, the growth of consumption in the U.S. has been stimulated partly by increasing demand for tropical and exotic fruits and vegetables.

Economic and Social Impact

Ongoing consumer demand for new fruits and vegetables in developed countries has contributed to an increase in trade volume of fresh produce in developing countries. This, in turn, has promoted the growth of small farms and the addition of new products, creating more rural and urban jobs and reduced the disparities in income levels among farms of different sizes. As countries become wealthier, their demand for high-valued commodities increases. The effect of income growth on consumption is more pronounced in developing countries, compared to developed countries, they are expected to spend larger shares of extra income on food items like meat and fruit and vegetable products. The implementation of international trade agreements, such as NAFTA (U.S., Mexico, Canada) and MERCOSUR (Argentina, Brazil, Paraguay and Uruguay), has significantly impacted the economy of the signatory countries by increasing the trade volumes and trade flows, particularly through general areas such as market access, tarification, limits on export subsidies, cuts in domestic supports, phyto-sanitary measures, and safeguard clauses.

Commercial Constraints

According to the USDA economic report, the commercial constraints on fruits and vegetables include:

— *Trade barriers:* Natural and artificial barriers. Natural trade barriers include high transportation costs to distant markets, and artificial barriers include legal measures such as protectionist policies. Liberalization of trade through international agreements has been instrumental in relaxing many legal trade barriers by reducing tariffs and by harmonizing the technical barriers to trade.

— *Scientific phyto-sanitary requirements:* Importing countries set the standards that potential trade partners must meet in order to protect human health or prevent the spread of pests and diseases. For instance, Japanese imports of U.S. apples are limited to Red and Golden Delicious apples from Washington and Oregon. The Japanese, who are mainly concerned with the spread of fire blight, impose rigorous and costly import requirements on the U.S. apple shippers. The apples must be subjected to a cold treatment and fumigation with methyl bromide before shipment to Japan, and three inspections of U.S. apple orchards during the production stage. Infestation by fruit flies, common in the tropics, is a major constraint to the production and export of tropical fruits.

— *Technological innovations:* Countries can increase their competitiveness and world market shares by providing higher quality products and promoting lower prices through technological innovations.

Trade liberalization, negotiated through the Uruguay Round Agreement (URA) (of the GATT and implemented under WTO), as well as through regional agreements, such as NAFTA and MERCOSUR, has expanded market access and provided strengthened mechanisms for combating non-tariff trade barriers such as scientifically unfounded phyto-sanitary restrictions.

Future prospects of fruits and vegetables exported from developing countries will largely depend on the growth of import demand, mostly in the developed countries. Developed countries are expected to diversify their consumption of fruits and vegetables. This will increase the concern about health and nutrition; the consumer's familiarity with more fruits and vegetables because of wider availability, increased travel, and improved communications will lead to an increase in the ratio of imports to domestic products.

Post-harvest Losses and Resource Under-utilization

Postharvest losses of fruits and vegetables are difficult to predict; the major agents producing deterioration are those attributed to physiological damage and combinations of several organisms. Flores described postharvest losses due to various causes as follows:

Food Losses after Harvesting

These include losses from technological origin such as deterioration by biological or microbiological agents, and mechanical damage.

Losses due to technological origin include: unfavourable climate, cultural practices, poor storage conditions, and inadequate handling during transportation, all of which can lead to accelerated product decay (e.g., tubers re-sprouting from bulbs and weight loss from product dehydration).

Physiological deterioration of fruits and vegetables refers to the aging of products during storage due to natural reactions. Deterioration caused by biochemical or chemical agents refers to reactions, of which intermediate and final products are undesirable. These can result in significant loss of nutritional value (i.e., rancidity and agrochemical contamination) and in many cases the whole fruit or vegetable is lost.

Deterioration by biological or microbiological agents refers to losses caused by insects, bacteria, moulds, yeasts, viruses, rodents, and other animals. When fruits and vegetables are gathered into boxes, crates, baskets, or trucks after harvesting, they may be subject to cross-contamination by spoilage microorganisms from other fruits and vegetables and from containers.

Most of the microorganisms present in fresh vegetables are saprophytes, such as *coryniforms,* lactic acid bacteria, spore-formers, *coliforms, micrococci*, and pseudomonas, derived from the soil, air, and water. *Pseudomonas* and the group of *Klebsiella-Enterobacter-Serratia* from the enterobacteriaceae are the most frequent. Fungi, including*Aureobasidium, Fusarium*, and *Alternaria*, are often present but in relatively lower numbers than bacteria. Due to the acidity of raw fruits, the primary spoilage organisms are fungi, predominantly moulds and yeasts, such as*Sacharomyces cerevisiae*, *Aspergillus niger, Penicillum spp., Byssochlamys fulva*, *B. nivea*, *Clostridium pasteurianum*, *Coletotrichum gloesporoides*, *Clostridium perfringes*, and *Lactobacillus spp*. Psychrotrophic bacteria are able to grow in vegetable products; some of them are *Erwina carotovora*, *Pseudomonas fluorescens, P. auriginosa, P. luteola, Bacillus species, Cytophaga jhonsonae, Xantomonas campestri*, and *Vibrio fluvialis*.

The existence of pathogenic bacteria in fresh fruit and vegetable products has been reported by Alzamora et al., which include *Listeria monocytogenes, Aeromonas hydrophila,* and *Escherichia coli* O157: H7. These bacteria are found in both fresh and minimally processed fruit and vegetable products. *Listeria monocytogenes* is able to survive and grow at refrigeration temperatures on many raw and processed vegetables, such as ready-to-eat fresh salad vegetables, including cabbage, celery, raisins, fennel, watercress, leek salad, asparagus, broccoli, cauliflower, lettuce, lettuce juice, minimally-processed lettuce, butterhead lettuce salad, broad-leaved and curly-leaved endive, fresh peeled hamlin oranges, and vacuum-packaged potatoes. *Aeromonas hydrophila* is a characteristic concern in vegetables;

it is a psychrotrophic and facultative anaerobe. *Aeromonas*strains are susceptible to disinfectants, including chlorine, although recovery of *Aeromonas* from chlorinated water has been reported. Challenging studies inoculating *A. hydrophila* in minimally processed fruit salads showed that *A. hydrophila* was able to grow at 5°C during the first 6 days, however, the pathogen decreased after 8 days of storage.. *E. coli O157H:7* has emerged as a highly significant food borne pathogen. The principal reservoir of E. *coli O157H:7* is believed to be the bovine gastrointestinal tract. Thus, contamination of associated food products with faeces is a significant risk factor, particularly if untreated contaminated water is consumed directly or used to wash uncooked foods.

Mechanical damage is caused by inappropriate methods used during harvesting, packaging, and inadequate transporting, which can lead to tissue wounds, abrasion, breakage, squeezing, and escape of fruits or vegetables. Mechanical damage may increase susceptibility to decay and growth of microorganisms. Some operations, such as washing, can reduce the microbial load; however, they may also help to distribute spoilage microorganisms and moisten surfaces enough to permit growth of microorganisms during holding periods. All methods of harvesting cause bruising and damage to the cellular and tissue structure, in which enzyme activity is greatly enhanced as cellular components are dislocated.

Besides the above issues, most post-harvest losses in developing countries occur during transport, handling, storage, and processing. Rough handling during preparation for market will increase bruising and mechanical damage, and limits the benefits of cooling.

By-products from fruit and vegetable processing are not wholly utilized in developing countries due to lack of machinery and infrastructure to process waste. The easiest way to dispose of by-products is to dump the waste or use it directly as animal feed. Waste materials such as leaves and tissues could be used in animal feed formulations and plant fertilizers.

In general, it is estimated that between 49 to 80% goes to consumers in the production of a particular commodity, and the difference is lost during the varied steps that comprise the harvest-consumption system.

Food Losses Due to Social and Economic Reasons

— *Policies*: This involves political conditions under which a technological solution is inappropriate or difficult to put in to practice, for example, lack of a clear policy capable of facilitating and encouraging utilization and administration of human, economic, technical, and scientific resources to prevent the deterioration of commodities.

— *Resources*: This is related to human, economic, and technical resources for developing programs aimed at prevention and reduction of post-harvest food losses.

— *Education*: This includes unknown knowledge of technical and scientific technologies associated with preservation, processing, packaging, transporting, and distribution of food products.

— *Services:* This refers to inefficient commercialization systems, and absent or inefficient government agencies in the production and marketing of commodities, as well as a lack of credit policies that address the needs of the country and participants.

— *Transportation:* This is a serious problem faced by fruit growers in developing countries, where vehicles used in transporting bulk raw fruits to markets are not equipped with good refrigeration systems. Raw fruits exposed to high temperatures during transportation soften in tissue and bruise easily, causing rapid microbial deterioration.

Pre-processing to Add Value

Rapid cooling of produce following harvest is essential for crops intended for transport in refrigerated ships, land vehicles, and containers not designed to handle the full load of field heat but capable of maintaining precooled produce at a selected carriage temperature. The selected method of cooling will depend greatly on the anticipated storage life of the commodity. Rapidly respiring commodities with short post-harvest life should be cooled immediately after harvest. Therefore, added value is achieved in precooling the produce immediately after harvest, which will restrict deterioration and maintain the produce in a condition acceptable to the consumer.Blanching of fruits as a pre-treatment method may also be applied before freezing and juicing, or in some cases, before dehydration. The fruit may be blanched either by exposure to near boiling water, steam, or hot air for 1 to 10 minutes. Blanching inactivates those enzyme systems that degrade flavour and colour and cause vitamin loss during subsequent processing and storage.

Pre-processing to Avoid Losses

Pre-processing of fruits and vegetables includes: blanching to inactivate enzymes and microorganisms, curing of root and tubers to extend shelf life, pre-treatment of produce with cold or high temperatures, and chemical preservatives to control pests after harvest. Storage of produce under controlled temperature and relative humidity conditions will extend its perishability and reduce decay. Packaging of produce in appropriate material enhances colour appearance and marketability.

Alternative Processing Methods for Fruits and Vegetables

A variety of alternative methods to preserve fruits and vegetables can be used in rural areas, such as fermentation, sun drying, osmotic dehydration, and refrigeration.

Fruits and vegetables can be pre-processed via scalding (blanching) to eliminate enzymes and microorganisms. Fermentation of fruits and vegetables is a preservation method used in rural areas, and due to the simplicity of the process, there is no need for sophisticated equipment; pickled produce, sauerkraut, and wine are examples of this process.

Cleaning and washing are often the only preservation treatments applied to minimally processed raw fruits and vegetables (MPRFV). As the first step in processing, cleaning is a form of separation concerned with removal of foreign materials like twigs, stalks, dirt, sand, soil, insects, pesticides, and fertilizer residues from fruits and vegetables, as well as from containers and equipment. The cleaning process also involves separation of light from heavy materials via gravity, flotation, picking, screening, dewatering, and others. Washing is usually done with chlorinated water. The MPRFV product is immersed in a bath in which bubbling is maintained by a jet of air. This turbulence permits one to eliminate practically all traces of air and foreign matter without bruising the product.

Water must be of optimal quality for washing MPRFV products, otherwise cross contamination may occur. According to Wiley R.C., three parameters are controlled in washing MPRFV fruits and vegetables:

1. Quantity of water used: 5-10 L/kg of product
2. Temperature of water: 4°C to cool the product
3. Concentration of active chlorine: 100 mg/L

Two examples of specially designed equipment used to wash fruits and vegetables include: 1) rotary drums used for cleaning apples, pears, peaches, potatoes, turnips, beets; high pressure water is sprayed over the product, which never comes in contact with dirty water, and 2) wire cylinder leafy vegetable washers, in which medium pressure sprays of fresh water are used for washing spinach, lettuce, parsley, and leeks.

In rural areas, fresh produce could be poured into plastic containers filled with tap water to remove the dirt from fruits and vegetables. The dirty water could be drained from the containers and refilled with chlorinated water for rewashing and disinfection of the fruit or vegetable. If electricity is available, fresh produce could be refrigerated until processed or distributed to retailers and markets.

Scalding or blanching in hot water

Fruits, fresh vegetables and root vegetable pieces are immersed in a bath containing hot water (or boiling water) for 1-10 minutes at 91-99°C, to reduce microbial levels, and partially reduce peroxidase and polyphenoloxydase (PPO) activity. The heating time will depend on the type of vegetable product processed Boiling water has been used to provide thermal inactivation of *L. monocytogenes* on celery leaves.

Cooling in trays

This operation is carried out in perforated metal trays through which cool air is passed in order to cool the product prior to packaging in sterile plastic bags, unless another process is to follow.

Sulphiting

During this operation, the fruit or vegetable pieces (or slices) are immersed in a solution of sodium bisulphite (200 ppm) to prevent undesirable changes in colour and any additional microbial and enzyme activity, and to retain a residual concentration of 100 ppm in the final product.

Sun drying and osmotic dehydration

In rural areas, dehydration is probably the most effective method to preserve fruits and vegetables. Fruit slices or vegetable pieces are spread over stainless metal trays or screens spaced 2-3 cm apart and sun dried. The dried fruit and vegetable products are then packaged in plastic bags, glass bottles, or cans, as with fruit slices (i.e., mango, papaya, peach, etc.) or milled flour (i.e., green plantain flour produced in rural areas of developing countries).

In osmotic dehydration and crystallization, the fruit is preserved by heating the product in sugar syrup, followed by washing and drying to reduce the sugar concentration at the fruit surface. Fruits are dried by direct or indirect sun drying, depending on the quality of the product obtained. The advantage of this method is the prevention of discoloration and browning of fruit produced by enzymatic reactions. Thus, the high concentration of sugar in the fruit produces a dehydrated product with good colouring, without the need of chemical preservatives such as sulphur dioxide.

Fermentation

This is another useful preservation process for fruit and vegetable products. For vegetables, the product is immersed into a sodium chloride solution, as in the case of cucumbers, green tomatoes, cauliflower, onions, and cabbage (sauerkraut). Composition of the salt (sodium chloride) is maintained at about 12% by weight so that active organisms during fermentation, such as Lactic acid bacteria, and the *Aerobacter* group, produce sufficient acid to prevent any food poisoning organisms from germinating. Fruits, on the other hand, can be preserved by fermenting the fruit pulp into wine, by preparing a solution of sugar and water and then inoculating it with a strain of *Saccharomyces cerevisiae*.

Pickles, sauerkraut and wine making

Slightly underripe cucumbers are selected and cleaned thoroughly with water, then size-graded prior to brining. For a large production of pickles, the fermentation process is carried out in circular wooden vats 2.5-4.5 m in diameter and 1.8-2.5 m deep. A small batch of pickles can be produced using appropriate plastic containers capable of holding 4-5 kg of cucumbers. After the cucumbers are put into the vats, a salt solution (approximately 10% by weight) is added. This concentration is maintained by adding further salt as needed by recirculating the solution to eliminate concentration gradients. Sugar is added if the cucumbers are low in sugar content to sustain the fermentation process. The fermentation process will end after 4-6 weeks, and the salt concentration will rise to 15%. Under these conditions, pickles will keep almost indefinitely. Care must be taken to ensure that the yeast scum on top of the vat does not destroy the lactic acid. This can be done by adding a layer of liquid paraffin on the surface of the pickling solution. After the fermentation process has ended, the pickles are soaked in hot water to remove excess salt, then size-graded and packed into glass jars with acetic acid in the form of vinegar.

Sauerkraut

Selected heads of cabbage are core-shredded and soaked in tap water with 2.5% (by weight) salt concentration and allowed to ferment. During the initial stages of fermentation, there is a rapid evolution of gas caused by *Leuconostoc mesenteroides;* this process imparts much of the pleasant flavour to the product. The next stage involves*Lactobacillus cucumeris* fermentation, resulting in an increase of lactic acid; and finally after approximately 5 days at 20-24°C, the third stage, involving a further group of lactic acid bacteria such as *Leuconostoc pentoaceticus*, which yields more lactic acid combined with acetic acid, ethyl alcohol, carbon dioxide, and mannitol. The fermentation process ends when the lactic acid production is approximately 1-2%. This can be tested by titration of the acid with a 0.1 N sodium hydroxide (NaOH) solution, using phenolphthalein (0.1% w/v) as colour indicator (i.e., 2-5 drops are added to the acid solution; colour will change from clear to pink and persists for 30 seconds). After the fermentation process, either the tank is sealed to exclude air or the product is then packed into glass jars or canned. It is then ready for consumption.

Wine making

Selected ripened fruits are transported to the farm where they are sorted, washed and macerated or chopped prior to pressing. In rural areas, juice is extracted from the fruit by squeezing (oranges, grapes, etc.) or pulped (mangoes, maracuyá, guava, etc.). The soluble solid content of the pulp is measured with a refractometer in °Brix. Soluble solids should be 25%, but if lower, it can be adjusted with sugar.

Clarification: Clarification of wines prior to bottling involves treatment with gelatine, albumin, isinglass, bentonite, potassium ferrocyanide or salts (the last two treatments are intended to reduce the level of soluble iron complexes, which would otherwise cause a darkening of the wine, but with fruit wine these are frequently inadequate. Alternative clarification procedures include chilling the wine prior to, or after, refining, and using microfiltration systems. A simple way to clarify wine is to add white gelatine (1 g per L of wine) to the fermented fruit solution, which is then allowed to stand in the refrigerator for 1 week, after which all of the suspended solids are precipitated and a clear transparent wine can be decanted from the top of the container. Following clarification, the wine will normally be flash pasteurized, hot-filled into bottles, or treated to give a residual SO_2 content (100 ppm).

The next stage is to add sodium bisulphite to the fruit juice (200 ppm), allowing it to stand for 2-3 hours. During this process, the unwanted yeast flora present in the fruit pulp is eliminated and the added inoculum can act freely in the fruit juice to produce the desired flavour or bouquet characteristic of fruit wine. Next, the yeast is added to the juice (1 g per kg of fruit juice, usually strains of *Saccharomyces cerevisiae* or bread making yeast). The fermentation process should be carried out anaerobically, that is in the absence of oxygen, to prevent development of other non-wine making bacteria, such as *Acetobacter spp,* which produces undesirable taste and flavour. The fermentation ends after 3 to 4 weeks at 22-25°C.

The final stage of processing involves the blending, sweetening and flavouring (if required), and stabilization of the wines. The blending process is done both to ensure consistency of product character and to reduce the strong aroma and flavour of certain wines. Although there is some preference for single wines, many are blended, especially with apple wine, which is relatively low in flavour. Wines can be sweetened using sugar or fruit juice, the latter also serving to increase the natural fruit content. In some cases, it is necessary to adjust the acidity of wine by adding an approved food-grade acid, such as citric or tartaric acid. In many rural areas, where these chemicals are not available, lemon juice can be used instead. For wine making in rural areas, the fermentation process is usually carried out in a large bottle (18-20 L), in which the ingredients are mixed with water. In order to keep the fermentation process under anaerobic conditions, a water-filled air-lock is fitted into a hollow cork or rubber stopper inside the mouth of the bottle.

Storage

Because sun dried and fermented fruit and vegetable products are stable, they can be stored at ambient temperatures or at low refrigeration temperatures, extending the shelf life for several months (6-12 months and beyond). Wine is stored in glass bottles and maintained at room temperature or it can be stored under refrigeration. Other fermented products such as sauerkraut and pickles are usually stored at room temperature.

Harvest Handling

Maturity Index for Fruits and Vegetables

The principles dictating at which stage of maturity a fruit or vegetable should be harvested are crucial to its subsequent storage and marketable life and quality. Post-harvest physiologists distinguish three stages in the life span of fruits and vegetables: maturation, ripening, and senescence. Maturation is indicative of the fruit being ready for harvest. At this point, the edible part of the fruit or vegetable is fully developed in size, although it may not be ready for immediate consumption. Ripening follows or overlaps maturation, rendering the produce edible, as indicated by taste. Senescence is the last stage, characterized by natural degradation of the fruit or vegetable, as in loss of texture, flavour, etc.

Skin colour

This factor is commonly applied to fruits, since skin colour changes as fruit ripens or matures. Some fruits exhibit no perceptible colour change during maturation, depending on the type of fruit or vegetable. Assessment of harvest maturity by skin colour depends on the judgment of the harvester, but colour charts are available for cultivars, such as apples, tomatoes, peaches, chilli peppers, etc.

Optical methods

Light transmission properties can be used to measure the degree of maturity of fruits. These methods are based on the chlorophyll content of the fruit, which is reduced during maturation. The fruit is exposed to a bright light, which is then switched off so that the fruit is in total darkness. Next, a sensor measures the amount of light emitted from the fruit, which is proportional to its chlorophyll content and thus its maturity.

Shape

The shape of fruit can change during maturation and can be used as a characteristic to determine harvest maturity. For instance, a banana becomes more rounded in cross-sections and less angular as it develops on the plant. Mangoes also change shape during maturation. As the mango matures on the tree the relationship between the shoulders of the fruit and the point at which the stalk is attached may change. The shoulders of immature mangoes slope away from the fruit stalk; however, on more mature mangoes the shoulders become level with the point of attachment, and with even more maturity the shoulders may be raised above this point.

Size

Changes in the size of a crop while growing are frequently used to determine the time of harvest. For example, partially mature cobs of *Zea mays saccharata* are marketed as sweet corn, while even less mature and thus smaller cobs are marketed as baby corn. For bananas, the width of individual fingers can be used to determine harvest maturity. Usually a finger is placed midway along the bunch and its maximum width is measured with callipers; this is referred to as the calliper grade.

Aroma

Most fruits synthesize volatile chemicals as they ripen. Such chemicals give fruit its characteristic odour and can be used to determine whether it is ripe or not. These doors may only be detectable by humans when a fruit is completely ripe, and therefore has limited use in commercial situations.

Fruit opening

Some fruits may develop toxic compounds during ripening, such as ackee tree fruit, which contains toxic levels of hypoglycine. The fruit splits when it is fully mature, revealing black seeds on yellow arils. At this stage, it has been shown to contain minimal amounts of hypoglycine or none at all. This creates a problem in marketing; because the fruit is so mature, it will have a very short post-harvest life.

Analysis of hypoglycine 'A' (hyp.) in ackee tree fruit revealed that the seed contained appreciable hyp. at all stages of maturity, at approximately 1000 ppm, while levels in the membrane mirrored those in the arils. This analysis supports earlier observations that unopened or partially opened ackee fruit should not be consumed, whereas fruit that opens naturally to over 15 mm of lobe separation poses little health hazard, provided the seed and membrane portions are removed.

These observations agree with those of Brown et al. who stated that bright red, full sized ackee should never be forced open for human consumption.

Leaf changes

Leaf quality often determines when fruits and vegetables should be harvested. In root crops, the condition of the leaves can likewise indicate the condition of the crop below ground. For example, if potatoes are to be stored, then the optimum harvest time is soon after the leaves and stems have died. If harvested earlier, the skins will be less resistant to harvesting and handling damage and more prone to storage diseases.

Abscission

As part of the natural development of a fruit an abscission layer is formed in the pedicel. For example, in cantaloupe melons, harvesting before the abscission layer is fully developed results in inferior flavoured fruit, compared to those left on the vine for the full period.

Firmness

A fruit may change in texture during maturation, especially during ripening when it may become rapidly softer. Excessive loss of moisture may also affect the texture of crops. These textural changes are detected by touch, and the harvester may simply be able to gently squeeze the fruit and judge whether the crop can be harvested. Today sophisticated devices have been developed to measure texture in fruits and vegetables, for example, texture analyzers and pressure testers; they are currently available for fruits and vegetables in various forms.

A force is applied to the surface of the fruit, allowing the probe of the penetrometer or texturometer to penetrate the fruit flesh, which then gives a reading on firmness. Hand held pressure testers could give variable results because the basis on which they are used to measure firmness is affected by the angle at which the force is applied. Two commonly used pressure testers to measure the firmness of fruits and vegetables are the Magness-Taylor and UC Fruit Firmness testers.

A more elaborate test, but not necessarily more effective, uses instruments like the Instron Universal Testing Machine. It is necessary to specify the instrument and all settings used when reporting test pressure values or attempting to set standards.

The Agricultural Code of California states that "Bartlett pears shall be considered mature if they comply with one of the following: (a) the average pressure test of not less than 10 representative pears for each commercial size in any lot does not exceed 23 lb (10.4 kg); (b) the soluble solids in a sample of juice from not less than 10 representative pears for each commercial size in any lot is not less than 13%". This Code defines minimum maturity for Bartlett pears and is presented in Table 1.

Table 1 can be simplified by establishing a minimum tolerance level of 13% soluble solids as indicator of a pear's maturity and in this way avoid the pressure test standard control.

Table 1. Minimum maturity standard (expressed as minimum soluble solids required and maximum Magness-Taylor test pressure allowed) of fresh Bartlett pears for selected pear size ranges

*Pear Size**	*6.0 cm to 6.35 cm*	*£ 6.35 cm*
Minimum Soluble Solids (%)	*Maximum Test Pressure (kg)*	
Below 10%	8.6	9.1
10%	9.1	9.5
11%	9.3	9.8
12%	9.5	10.0

**Pear size expressed as maximum diameter (cm)*

Juice content

The juice content of many fruits increases as the fruit matures on the tree. To measure the juice content of a fruit, a representative sample of fruit is taken and then the juice extracted in a standard and specified manner. The juice volume is related to the original mass of juice, which is proportional to its maturity. The minimum values for citrus juices are presented in Table 2.

Table 2. Minimum juice values for mature citrus.

Citrus fruit	*Minimum juice content (%)*
Naval oranges	30
Other oranges	35
Grapefruit	35
Lemons	25
Mandarins	33
Clementines	40

Oil content and dry matter percentage

Oil content can be used to determine the maturity of fruits, such as avocados. According to the Agricultural Code in California, avocados at the time of harvest and at any time thereafter, shall not contain in weight less than 8% oil per avocado, excluding skin and seed. Thus, the oil content of an avocado is related to moisture content. The oil content is determined by weighing 5-10 g of avocado pulp and then extracting the oil with a solvent in a destillation column. This method has been successful for cultivars naturally high in oil content.

A round flask is used for the solvent. Heat is supplied with an electric plate and water recirculated to maintain a constant temperature during the extraction process. Extraction is performed using solvents such as petroleum ether, benzene, diethyl ether, etc., a process that takes between 4-6 h. After the extraction, the oil is recovered from the flask through evaporation of the water at 105°C in an oven until constant weight is achieved.

Moisture content

During the development of avocado fruit the oil content increases and moisture content rapidly decreases. The moisture levels required to obtain good acceptability of a variety of avocados cultivated in Chile are listed in Table 3.

Table 3. Moisture content of avocado fruit cultivated in Chile.

Cultivar	*Moisture content* (%)
Negra de la Cruz	80.1
Bacon	77.5
Zutano	80.5
Fuerte	77.9
Edranol	78.1
Hass	73.8
Gwen	78.4
Whitesell	79.1

Sugars

In climacteric fruits, carbohydrates accumulate during maturation in the form of starch. As the fruit ripens, starch is broken down into sugar. In non-climacteric fruits, sugar tends to accumulate during maturation. A quick method to measure the amount of sugar present in fruits is with a brix hydrometer or a refractometer.

A drop of fruit juice is placed in the sample holder of the refractometer and a reading taken; this is equivalent to the total amount of soluble solids or sugar content. This factor is used in many parts of the world to specify maturity. The soluble solids content of fruit is also determined by shining light on the fruit or vegetable and measuring the amount transmitted. This is a laboratory technique however and might not be suitable for village level production.

Starch content

Measurement of starch content is a reliable technique used to determine maturity in pear cultivars. The method involves cutting the fruit in two and dipping the cut pieces into a solution containing 4% potassium iodide and 1% iodine. The cut surfaces stain to a blue-black colour in places where starch is present. Starch converts into sugar as harvest time approaches. Harvest begins when the samples show that 65-70% of the cut surfaces have turned blue-black.

Acidity

In many fruits, the acidity changes during maturation and ripening, and in the case of citrus and other fruits, acidity reduces progressively as the fruit matures on the tree.

Taking samples of such fruits, and extracting the juice and titrating it against a standard alkaline solution, gives a measure that can be related to optimum times of harvest. Normally, acidity is not taken as a measurement of fruit maturity by itself but in relation to soluble solids, giving what is termed the brix: acid ratio. Sanchez et al. studied the effect of inducing maturity in banana (*Musa sp (L.), AAB*) "Silk" fruits with 2-chloroethyl phosphoric acid ("ethephon"), in some trials in Venezuela. Four treatments (0, 1000, 3000, and 5000 ppm) were applied. The results obtained revealed that the "ethephon" treatments increased the acidity and total soluble solids. The sucrose formation accelerated while the pH was not affected significantly.

Specific gravity

Specific gravity is the relative gravity, or weight of solids or liquids, compared to pure distilled water at 62°F (16.7°C), which is considered unity. Specific gravity is obtained by comparing the weights of equal bulks of other bodies with the weight of water. In practice, the fruit or vegetable is weighed in air, then in pure water. The weight in air divided by the weight in water gives the specific gravity. This will ensure a reliable measure of fruit maturity. As a fruit matures its specific gravity increases. This parameter is rarely used in practice to determine time of harvest, but could be used in cases where development of a suitable sampling technique is possible. It is used however to grade crops according to different maturities at post-harvest. This is done by placing the fruit in a tank of water, wherein those that float are less mature than those that sink.

Harvesting Containers

Harvesting containers must be easy to handle for workers picking fruits and vegetables in the field. Many crops are harvested into bags. Harvesting bags with shoulder or waist slings can be used for fruits with firm skins, like citrus fruits and avocados. These containers are made from a variety of materials such as paper, polyethylene film, sisal, hessian or woven polyethylene and are relatively cheap but give little protection to the crop against handling and transport damage. Sacks are commonly used for crops such as potatoes, onions, cassava, and pumpkins. Other types of field harvest containers include baskets, buckets, carts, and plastic crates. For high risk products, woven baskets and sacks are not recommended because of the risk of contamination.

Tools for Harvesting

Depending on the type of fruit or vegetable, several devices are employed to harvest produce. Commonly used tools for fruit and vegetable harvesting are secateurs or knives, and hand held or pole mounted picking shears. When fruits or vegetables are difficult to catch, such as mangoes or avocados, a cushioning material is placed around the tree to prevent damage to the fruit when dropping from high trees. Harvesting bags with shoulder or waist slings can be used for fruits with firm skins, like citrus and avocados. They are easy to carry and leave both hands free. The contents of the bag are emptied through the bottom into a field container without tipping the bag. Plastic buckets are suitable containers for harvesting fruits that are easily crushed, such as tomatoes. These containers should be smooth without any sharp edges that could damage the produce. Commercial growers use bulk bins with a capacity of 250-500 kg, in which crops such as apples and cabbages are placed, and sent to large-scale packinghouses for selection, grading, and packing.

Packing in the Field and Transport to Packinghouse

Berries picked for the fresh market are often mechanically harvested and usually packed into shipping containers. Careful harvesting, handling, and transporting of fruits and vegetables to packinghouses are necessary to preserve product quality.

Polyethylene bags

Clear polyethylene bags are used to pack banana bunches in the field, which are then transported to the packinghouse by means of mechanical cableways running through the banana plantation. This technique of packaging and transporting bananas reduces damage to the fruit caused by improper handling.

Plastic field boxes

These types of boxes are usually made of polyvinyl chloride, polypropylene, or polyethylene. They are durable and can last many years. Many are designed in such a way that they can nest inside each other when empty to facilitate transport, and can stack one on top of the other without crushing the fruit when full.

Wooden field boxes

These boxes are made of thin pieces of wood bound together with wire. They come in two sizes: the bushel box with a volume of 2200 in^3 (36052 cm^3) and the half-bushel box. They are advantageous because they can be packed flat and are inexpensive, and thus could be non-returnable. They have the disadvantage of providing little protection from mechanical damage to the produce during transport. Rigid wooden boxes of different capacities are commonly used to transport produce to the packinghouse or to market.

Bulk bins

Bulk bins of 200-500 kg capacity are used for harvesting fresh fruits and vegetables. These bins are much more economical than the field boxes, both in terms of fruit carried per unit volume and durability, as well as in providing better protection to the product during transport to the packinghouse. They are made of wood and plastic materials. Dimensions for these bins in the United States are 48 × 40 in, and 120 × 100 cm in metric system countries. Approximate depth of bulk bins depends on the type of fruit or vegetable being transported.

Post-harvest Handling

Curing of roots, tubers, and bulb crops

When roots and tubers are to be stored for long periods, curing is necessary to extend the shelf life. The curing process involves the application of high temperatures and high relative humidity to the roots and tubers for long periods, in order to heal the skins wounded during harvesting. With this process a new protected layer of cells is formed. Initially the curing process is expensive, but in the long run, it is worthwhile.

Curing can be accomplished in the field or in curing structures conditioned for that purpose. Commodities such as yams can be cured in the field by piling them in a partially shaded area. Cut grass or straw can serve as insulating material while covering the pile with canvas, burlap, or woven grass matting. This covering will provide sufficient heat to reach high temperatures and high relative humidity. The stack can be left in this state for up to four days.

Onions and garlic can be cured in the field in windrows or after being packed into large fibre or net sacks. Modern curing systems have been implemented in housing conditioned with fans and heaters to produce the heat necessary for high temperatures and high relative humidity, as illustrated below:

The fans are used to redistribute the heat to the lower part of the room where the produce is stored. Bulk bins are stacked with a gap of 10 to 15 cm between rows to allow adequate air passage. Care should be taken to prevent over-dryness of the onion bulbs.

When extreme conditions in the field exist, such as heavy rain or flooded terrain, and curing facilities are not available, a temporary tent must be constructed from large tarpaulins or plastic sheets to cure the onions and avoid heavy loss. Heated air is forced into a hollow area at the centre of the produce-filled bins. Several fans are used to recirculate the warm air through the onions while curing.

Operations prior to packaging

Fruits and vegetables are subjected to preliminary treatments designed to improve appearance and maintain quality. These preparatory treatments include cleaning, disinfection, waxing, and adding of colour (some includes brand name stamping on individual fruits).

Cleaning

Most produce receives various chemical treatments such as spraying of insecticides and pesticides in the field. Most of these chemicals are poisonous to humans, even in small concentrations. Therefore, all traces of chemicals must be removed from produce before packing. The fruit or vegetable passes over rotary brushes where it is rotated and transported to the washing machine and exposed to the cleaning process from all sides:

From the washing machine, the fruit passes onto a set of rotary sponge rollers (similar to the rotary brushes). The rotary sponges remove most of the water on the fruit as it is rotated and transported through the sponger.

Disinfection

After washing fruits and vegetables, disinfectant agents are added to the soaking tank to avoid propagation of diseases among consecutive batches of produce. In a soaking tank, a typical solution for citrus fruit includes a mixture of various chemicals at specific concentration, pH, and temperature, as well as detergents and water softeners. Sodium-ortho-phenyl-phenate (SOPP) is an effective citrus disinfectant, but requires precise control of conditions in the tank. Concentrations must be kept between 0.05 and 0.15%, with pH at 11.8 and temperature in the range of 43-48°C. Recommended soaking time is 3-5 minutes. Deviation from these recommendations may have disastrous effects on the produce, since the solution will be ineffective if the temperature or concentration is too low. Low concentrations of chlorine solution are also used as disinfectant for many vegetables. The advantage of this solution is that it does not leave a chemical residue on the product.

Artificial waxing

Artificial wax is applied to produce to replace the natural wax lost during washing of fruits or vegetables. This adds a bright sheen to the product. The function of artificial waxing of produce is summarized below:

— Provides a protective coating over entire surface.

— Seals small cracks and dents in the rind or skin.

— Seals off stem scars or base of petiole.

— Reduces moisture loss.

— Permits natural respiration.

— Extends shelf life.

— Enhances sales appeal.

Brand name application:

Some distributors use ink or stickers to stamp a brand name or logo on each individual fruit. Ink is not permissible in some countries (e.g., Japan), but stickers are acceptable. Automatic machines for dispensing and applying pressure sensitive paper stickers are readily available. The advantage of stickers is that they can be easily peeled off.

Packaging

According to Wills et al., modern packaging must comply with the following requirements:

a) The package must have sufficient mechanical strength to protect the contents during handling, transport, and stacking.

b) The packaging material must be free of chemical substances that could transfer to the produce and become toxic to man.

c) The package must meet handling and marketing requirements in terms of weight, size, and shape.

d) The package should allow rapid cooling of the contents. Furthermore, the permeability of plastic films to respiratory gases could also be important.

e) Mechanical strength of the package should be largely unaffected by moisture content (when wet) or high humidity conditions.

f) The security of the package or ease of opening and closing might be important in some marketing situations.

g) The package must either exclude light or be transparent.

h) The package should be appropriate for retail presentations.

i) The package should be designed for ease of disposal, re-use, or recycling.

j) Cost of the package in relation to value and the extent of contents protection required should be as low as possible.

Classification of packaging

Packages can be classified as follows:

— Flexible sacks; made of plastic jute, such as bags (small sacks) and nets (made of open mesh)

— Wooden crates

— Cartons (fibreboard boxes)

— Plastic crates

— Pallet boxes and shipping containers

— Baskets made of woven strips of leaves, bamboo, plastic, etc.

Uses for above packages

Nets are only suitable for hard produce such as coconuts and root crops (potatoes, onions, yams).

Wooden crates are typically wire bound crates used for citrus fruits and potatoes, or wooden field crates used for softer produce like tomatoes. Wooden crates are resistant to weather and more efficient for large fruits, such as watermelons and other melons, and generally have good ventilation. Disadvantages are that rough surfaces and splinters can cause damage to the produce, they can retain undesirable odours when painted, and raw wood can easily become contaminated with moulds.

Fibreboard boxes are used for tomato, cucumber, and ginger transport. They are easy to handle, light weight, come in different sizes, and come in a variety of colours that can make produce more attractive to consumers. They have some disadvantages, such as the effect of high humidity, which can weaken the box; neither are they waterproof, so wet products would need to be dried before packaging. These boxes are often of lower strength compared to wooden or plastic crates, although multiple thickness trays are very widely used. They can come flat packed with ventilation holes and grab handles, making a cheap attractive alternative that is very popular. Care should be taken that holes on the surface (top and sides) of the box allow adequate ventilation for the produce and prevent heat generation, which can cause rapid product deterioration.

Plastic crates are expensive but last longer than wooden or carton crates. They are easy to clean due to their smooth surface and are hard in strength, giving protection to products. Plastic crates can be used many times, reducing the cost of transport. They are available in different sizes and colours and are resistant to adverse weather conditions. However, plastic crates can damage some soft produce due to their hard surfaces, thus liners are recommended when using such crates.

Pallet boxes are very efficient for transporting produce from the field to the packinghouse or for handling produce in the packinghouse. Pallet boxes have a standard floor size (1200 × 1000 mm) and depending on the commodity have standard heights. Advantages of the pallet box are that it reduces the labour and cost of loading, filling, and unloading; reduces space for storage; and increases speed of mechanical harvest. The major disadvantage is that the return volume of most pallet boxes is the same as the full load. Higher investment is also required for the forklift truck, trailer, and handling systems to empty the boxes. They are not affordable to small producers because of high, initial capital investment.

Cooling methods and temperatures

Several methods of cooling are applied to produce after harvesting to extend shelf life and maintain a fresh-like quality. Some of the low temperature treatments are unsuitable for simple rural or village treatment but are included for consideration as follows:

Precooling

Fruit is precooled when its temperature is reduced from 3 to 6°C (5 to 10°F) and is cool enough for safe transport. Precooling may be done with cold air, cold water (hydrocooling), direct contact with ice, or by evaporation of water from the product under a partial vacuum (vacuum cooling). A combination of cooled air and water in the form of a mist called hyraircooling is an innovation in cooling of vegetables.

Air precooling

Precooling of fruits with cold air is the most common practice. It can be done in refrigerator cars, storage rooms, tunnels, or forced air-coolers (air is forced to pass through the container via baffles and pressure differences).

Icing

Ice is commonly added to boxes of produce by placing a layer of crushed ice directly on the top of the crop. An ice slurry can be applied in the following proportion: 60% finely crushed ice, 40% water, and 0.1% sodium chloride to lower the melting point. The water to ice ratio may vary from 1:1 to 1:4.

Room cooling

This method involves placing the crop in cold storage. The type of room used may vary, but generally consists of a refrigeration unit in which cold air is passed through a fan. The circulation may be such that air is blown across the top of the room and falls through the crop by convection. The main advantage is cost because no specific facility is required.

Forced air-cooling

The principle behind this type of precooling is to place the crop into a room where cold air is directed through the crop after flowing over various refrigerated metal coils or pipes. Forced air-cooling systems blow air at a high velocity leading to desiccation of the crop. To minimize this effect, various methods of humidifying the cooling air have been designed such as blowing the air through cold water sprays.

Hydrocooling

The transmission of heat from a solid to a liquid is faster than the transmission of heat from a solid to a gas. Therefore, cooling of crops with cooled water can occur quickly and results in zero loss of weight. To achieve high performance, the crop is submerged in cold water, which is constantly circulated through a heat exchanger. When crops are transported around the packhouse in water, the transport can incorporate a hydrocooler. This system has the advantage wherein the speed of the conveyer can be adjusted to the time required to cool the produce. Hydrocooling has a further advantage over other precooling methods in that it can help clean the produce. Chlorinated water can be used to avoid spoilage of the crop. Hydrocooling is commonly used for vegetables, such as asparagus, celery, sweet corn, radishes, and carrots, but it is seldom used for fruits.

Vacuum cooling

Cooling in this case is achieved with the latent heat of vaporization rather than conduction. At normal air pressure (760 mmHg) water will boil at 100°C. As air pressure is reduced so is the boiling point of water, and at 4.6 mmHg water boils at 0°C. For every 5 or 6°C reduction in temperature, under these conditions, the crop loses about 1% of its weight. This weight loss may be minimized by spraying the produce with water either before enclosing it in the vacuum chamber or towards the end of the vacuum cooling operation (hydrovacuum cooling). The speed and effectiveness of cooling is related to the ratio between the mass of the crop and its surface area. This method is particularly suitable for leaf crops such as lettuce. Crops like tomatoes having a relatively thick wax cuticle are not suitable for vacuum cooling.

Recommended minimum temperature to increase storage time

There is no ideal storage for all fruits and vegetables, because their response to reduced temperatures varies widely. The importance of factors such as mould growth and chilling injuries must be taken into account, as well as the required length of storage. Storage temperature for fruits and vegetables can range from -1 to 13°C, depending on their perishability. Extremely perishable fruits such as apricots, berries, cherries, figs, watermelons can be stored at -1 to 4°C for 1-5 weeks; less perishable fruits such as mandarin, nectarine, ripe or green pineapple can be stored at 5-9°C for 2-5 weeks; bananas at 10°C for 1-2 weeks and green bananas at 13°C for 1-2 weeks. Highly perishable vegetables can be stored up to 4 weeks such as asparagus, beans, broccoli, and Brussels sprouts at -1-4°C for 1-4 weeks; cauliflower at 5-9°C for 2-4 weeks. Green tomato is less perishable and can be stored at 10°C for 3-6 weeks and non-perishable vegetables such as carrots, onions, potatoes and parsnips can be stored at 5-9°C for 12-28 weeks. Similarly, sweet potatoes can be stored at 10°C for 16-24 weeks. The storage life of produce is highly variable and related to the respiration rate; there is an inverse relation between respiration rate and storage life in that produce with low respiration generally keeps longer.

For example, the respiration rate of a very perishable fruit like ripe banana is 200 mL $CO_2.kg^{-1}h^{-1}$ at 15°C, compared to a non-perishable fruit such as apple, which has a respiration rate of 25 mL CO_2.kg-1 h-1 at 15°C.

High temperatures

Exposure of fruits and vegetables to high temperatures during post-harvest reduces their storage or marketable life. This is because as living material, their metabolic rate is normally higher with higher temperatures. High temperature treatments are beneficial in curing root crops, drying bulb crops, and controlling diseases and pests in some fruits. Many fruits are exposed to high temperatures in combination with ethylene (or another suitable gas) to initiate or improve ripening or skin colour.

Storage

The marketable life of most fresh vegetables can be extended by prompt storage in an environment that maintains product quality. The desired environment can be obtained in facilities where temperature, air circulation, relative humidity, and sometimes atmosphere composition can be controlled. Storage rooms can be grouped accordingly as those requiring refrigeration and those that do not. Storage rooms and methods not requiring refrigeration include: *in situ*, sand, coir, pits, clamps, windbreaks, cellars, barns, evaporative cooling, and night ventilation:

In situ. This method of storing fruits and vegetables involves delaying the harvest until the crop is required. It can be used in some cases with root crops, such as cassava, but means that the land on which the crop was grown will remain occupied and a new crop cannot be planted. In colder climates, the crop may be exposed to freezing and chilling injury.

Sand or coir: This storage technique is used in countries like India to store potatoes for longer periods of time, which involves covering the commodity under ground with sand.

Pits or trenches are dug at the edges of the field where the crop has been grown. Usually pits are placed at the highest point in the field, especially in regions of high rainfall. The pit or trench is lined with straw or other organic material and filled with the crop being stored, then covered with a layer of organic material followed by a layer of soil. Holes are created with straw at the top to allow for air ventilation, as lack of ventilation may cause problems with rotting of the crop.

Clamps. This has been a traditional method for storing potatoes in some parts of the world, such as Great Britain. A common design uses an area of land at the side of the field. The width of the clamp is about 1 to 2.5 m. The dimensions are marked out and the potatoes piled on the ground in an elongated conical heap. Sometimes straw is laid on the soil before the potatoes. The central height of the heap depends on its angle of repose, which is about one third the width of the clump. At the top, straw is bent over the ridge so that rain will tend to run off the structure. Straw thickness should be from 15-25 cm when compressed. After two weeks, the clamp is covered with soil to a depth of 15-20 cm, but this may vary depending on the climate.

Windbreaks are constructed by driving wooden stakes into the ground in two parallel rows about 1 m apart. A wooden platform is built between the stakes about 30 cm from the ground, often made from wooden boxes. Chicken wire is affixed between the stakes and across both ends of the windbreak. This method is used in Britain to store onions.

Cellars. These underground or partly underground rooms are often beneath a house. This location has good insulation, providing cooling in warm ambient conditions and protection from excessively low temperatures in cold climates. Cellars have traditionally been used at domestic scale in Britain to store apples, cabbages, onions, and potatoes during winter.

Barns. A barn is a farm building for sheltering, processing, and storing agricultural products, animals, and implements. Although there is no precise scale or measure for the type or size of the building, the term barn is usually reserved for the largest or most important structure on any particular farm. Smaller or minor agricultural buildings are often labelled sheds or outbuildings and are normally used to house smaller implements or activities.

Evaporative cooling. When water evaporates from the liquid phase into the vapour phase energy is required. This principle can be used to cool stores by first passing the air introduced into the storage room through a pad of water. The degree of cooling depends on the original humidity of the air and the efficiency of the evaporating surface. If the ambient air has low humidity and is humidified to around 100% RH, then a large reduction in temperature will be achieved. This can provide cool moist conditions during storage.

Night ventilation. In hot climates, the variation between day and night temperatures can be used to keep stores cool. The storage room should be well insulated when the crop is placed inside. A fan is built into the store room, which is switched on when the outside temperature at night becomes lower than the temperature within. The fan switches off when the temperatures equalize. The fan is controlled by a differential thermostat, which constantly compares the outside air temperature with the internal storage temperature. This method is used to store bulk onions.

Controlled atmospheres are made of gastight chambers with insulated walls, ceiling, and floor. They are increasingly common for fruit storage at larger scale. Depending on the species and variety, various blends of O_2, CO_2, and N_2 are required. Low content O_2 atmospheres (0.8 to 1.5%), called ULO (Ultra -Low Oxygen) atmospheres, are used for fruits with long storage lives (e.g., apples).

Pest control and decay

Crops may be immersed in hot water before storage or marketing to control disease. A common disease of fruits known as anthracnose, caused by the infection of fungus *Colletotrychum spp*. can be successfully controlled in this way. Combining appropriate doses of fungicides with hot water is often effective in controlling disease in fruits after harvesting.

Fruit and vegetable decay is also caused by storage conditions. Too low temperatures can cause injury during refrigeration of fruits and vegetables. High temperatures can cause softening of tissues and promote bacterial diseases. The damage that microorganisms inflict on fresh fruits and vegetables is mainly in the physical loss of edible matter, which may be partial or total.

Water Activity (a_w) Concept and Its Role in Food Preservation

The concept of a_w has been very useful in food preservation and on that basis many processes could be successfully adapted and new products designed. Water has been called the universal solvent as it is a requirement for growth, metabolism, and support of many chemical reactions occurring in food products. Free water in fruit or vegetables is the water available for chemical reactions, to support microbial growth, and to act as a transporting medium for compounds. In the bound state, water is not available to participate in these reactions as it is bound by water soluble compounds such as sugar, salt, gums, etc. (osmotic binding), and by the surface effect of the substrate (matrix binding). These water-binding effects reduce the vapour pressure of the food substrate according to Raoult's Law. Comparing this vapour pressure with that of pure water (at the same temperature) results in a ratio called water activity (a_w). Pure water has an a_w of 1, one molal solution of sugar - 0.98, and one molal solution of sodium chloride - 0.9669. A saturated solution of sodium chloride has a water activity of 0.755. This same NaCl solution in a closed container will develop an equilibrium relative humidity (ERH) in a head space of 75.5%. A relationship therefore exists between ERH and a_w since both are based on vapour pressure.

$$a_w = \frac{\text{ERH}}{100}$$

The ERH of a food product is defined as the relative humidity of the air surrounding the food at which the product neither gains nor loses its natural moisture and is in equilibrium with the environment.

Microorganisms vs a_w Value

The definition of moisture conditions in which pathogenic or spoilage microorganisms cannot grow is of paramount importance to food preservation. It is well known that each microorganism has a critical a_w below which growth cannot occur. For instance, pathogenic microorganisms cannot grow at a_w <0.86; yeasts and moulds are more tolerant and usually no growth occurs at a_w <0.62. The so-called intermediate moisture foods (IMF) have a_w values in the range of 0.65-0.90 (Figure 1).

Enzymatic and Chemical Changes Related to a_w Values

The relationship between enzymatic and chemical changes in foods as a function of water activity is illustrated in Figure 1. With a_w at 0.3, the product is most stable with respect to lipid oxidation, non-enzymatic browning, enzyme activity, and of course, the various microbial parameters. As a_w increases toward the right, the probability of the food product deteriorating increases.

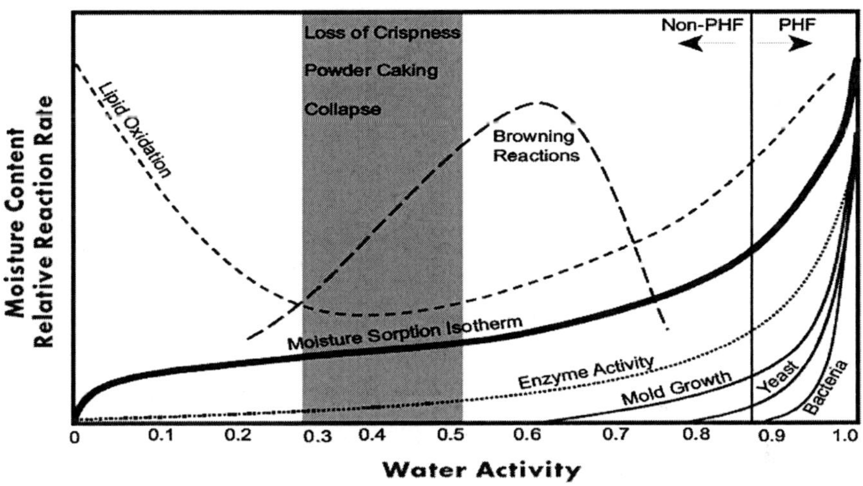

Figure 1. Relationship of food deterioration rate as a function of water activity

According to Rahman and Labuza, enzyme-catalyzed reactions can occur in foods with relatively low water contents. The authors summarized two features of these results as follows:

1. The rate of hydrolysis increases with increased water activity but is extremely slow with very low activity.
2. For each instance of water activity there appears to be a maximum amount of hydrolysis, which also increases with water content.

The apparent cessation of the reaction at low moisture cannot be due to the irreversible inactivation of the enzyme, because upon humidification to a higher water activity, hydrolysis resumes at a rate characteristic of the newly attained water activity. Rahman and Labuza reported the investigation of a model system consisting of avicel, sucrose, and invertase and found that the reaction velocity increased with water activity. Complete conversion of the substrate was observed for water activities greater than or equal to 0.75. For water activities below 0.75, the reaction continued with 100% hydrolysis. In solid media, water activity can affect reactions in two ways: lack of reactant mobility and alternation of active conformation of the substrate and enzymatic protein. The effects of varying the enzyme-to-substrate ratios on reaction velocity and the effect of water activity on the activation energy for the reaction could not be explained by a simple diffusional model, but required postulates that were more complex:

1. The diffusional resistance is localized in a shell adjacent to the enzyme.
2. At low water activity, the reduced hydration produces conformational changes in the enzyme, affecting its catalytic activity.

The relationship between water content and water activity is complex. An increase in aw is usually accompanied by an increase in water content, but in a non-linear fashion. This relationship between water activity and moisture content at a given temperature is called the moisture sorption isotherm. These curves are determined experimentally and constitute the fingerprint of a food system.

Recommended Equipment for Measuring a_w

Many methods and instruments are available for laboratory measurement of water activity in foods. Methods are based on the colligative properties of solutions. Water activity can be estimated by measuring the following:

— Vapour pressure
— Osmotic pressure
— Freezing point depression of a liquid
— Equilibrium relative humidity of a liquid or solid
— Boiling point elevation
— Dew point and wet bulb depression
— Suction potential, or by using the isopiestic method
— Bithermal equilibrium
— Electric hygrometers
— Hair hygrometers

Vapour pressure

Water activity is expressed as the ratio of the partial pressure of water in a food to the vapour pressure of pure water with the same temperature as the food. Thus, measuring the vapour pressure of water in a food system is the most direct measure of a_w. The food sample measured is allowed to equilibrate, and measurement is taken by using a manometer or transducer device as depicted in Figure 2. This method can be affected by sample size, equilibration time, temperature, and volume. This method is not suitable for biological materials with active respiration or materials containing large amounts of volatiles.

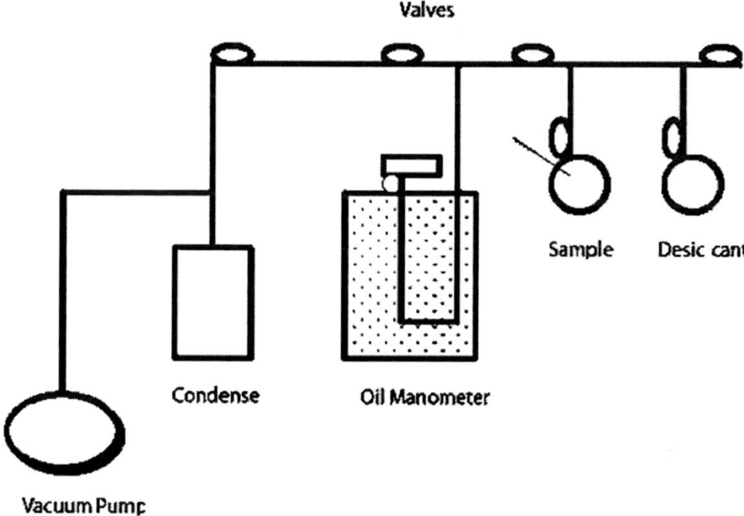

Figure 2. Vapour pressure manometer

Freezing point depression and freezing point elevation

This method is accurate for liquids in the high water activity range but is not suitable for solid foods. The water activity can be estimated using the following two expressions:

Freezing point depression:

$$-\log a_w = 0.004207\ DT_f + 2.1\ E\text{-}6\ DT^2_f \quad (1)$$

where DT_f is the depression in the freezing temperature of water,

Boiling point elevation:

$$-\log a_w = 0.01526\ DT_b - 4.862\ E\text{-}5\ DT^2_b \quad (2)$$

where DT_b is the elevation in the boiling temperature of water.

Osmotic pressure

Water activity can be related to the osmotic pressure (p) of a solution with the following equation:

$$p = RT/V_w \ln(a_w) \quad (3)$$

where V_w is the molar volume of water in solution, R the universal gas constant, and T the absolute temperature. Osmotic pressure is defined as the mechanical pressure needed to prevent a net flow of solvent across a semi-permeable membrane. For an ideal solution, Equation (3) can be redefined as:

$$p = RT/V_w \ln(X_w) \quad (4)$$

where X_w is the molar fraction of water in the solution. For non-ideal solutions, the osmotic pressure expression can be rewritten as:

$$p = RTfnm_b(m_w V_w) \quad (5)$$

where n is the number of moles of ions formed from one mole of electrolyte, mw and mb are the molar concentrations of water and the solute, respectively, and f the osmotic coefficient, defined as:

$$f = -m_w \ln(a_w)/nmb \quad (6)$$

Dew point hygrometer

Vapour pressure can be determined from the dew point of an air-water mixture. The temperature at which the dew point occurs is determined by observing condensation on a smooth, cool surface such as a mirror. This temperature can be related to vapour pressure using a psychrometric chart. The formation of dew is detected photoelectrically.

Thermocouple Psychrometer

Water activity measurement is based on wet bulb temperature depression.

A thermocouple is placed in the chamber where the sample is equilibrated. Water is then sprayed over the thermocouple before it is allowed to evaporate, causing a decrease in temperature. The drop in temperature is related to the rate of water evaporation from the surface of the thermocouple, which is a function of the relative humidity in equilibrium with the sample.

Isopiestic method

The isopiestic method consists of equilibrating both a sample and a reference material in an evacuated desiccator until equilibrium is reached at 25°C. The moisture content of the reference material is then determined and the a_w obtained from the sorption isotherm. Since the sample was in equilibrium with the reference material, the a_w of both is the same.

Electric hygrometers

Most hygrometers are electrical wires coated with hygroscopic salts or sulfonated polystyrene gel in which conductance or capacitance changes as the coating absorbs moisture from the sample. The major disadvantage of this type of hygrometer is the tendency of the hygroscopic salt to become contaminated with polar compounds, resulting in erroneous a_w determinations.

Hair hygrometers

Hair hygrometers are based on the stretching of a fibre when exposed to high water activity. They are less sensitive than other instruments at lower levels of activity (<0.03 a_w) and the principal disadvantage of these types of meters is the time delay in reaching equilibrium and the tendency to hysteresis.

Today we find many brands of water activity meters in the market. Most of these meters are based on the relationship between ERH and the food system, but differ in their internal components and configuration of software used. One of the water activity meters most used today is the AcquaLab Series 3 Model TE, developed by Decagon Devices, which is based on the chilled-mirror dew point method. This instrument is a temperature controlled water activity meter that allows placement of the sample in a temperature stable environment without the use of an external water bath. The temperature can be selected on the screen and is monitored and controlled with thermoelectric components. Most of the older generations of water activity instruments are based on a temperature-controlled environment. Therefore, a margin of error greater than 5% can be expected due to temperature variations. This equipment is highly recommended for measuring water activity in fruits and vegetables since it measures a wide range of water activity.

The major advantages of the chilled-mirror dew point method are accuracy, speed, ease of use and precision. The AquaLab's range is from 0.030 to 1.000a_w, with a resolution of ±0.001a_w and accuracy of ±0.003a_w. Measurement time is typically less than five minutes. Capacitance sensors have the advantage of being inexpensive, but are not usually as accurate or as fast as the chilled-mirror dew point method. Capacitive instruments measure over the entire water activity range 0 to 1.00 a_w, with a resolution of ±0.005a_w and accuracy of ±0.015a_w. Some commercial instruments can complete measurements in five minutes while other electronic capacitive sensors usually require 30 to 90 minutes to reach equilibrium relative humidity conditions.

Intermediate Moisture Foods (IMF) concept

Traditional intermediate moisture foods (IMF) can be regarded as one of the oldest foods preserved by man. The mixing of ingredients to achieve a given aw, that allowed safe storage while maintaining enough water for palatability, was only done, however, on an empirical basis. The work done by food scientists approximately three decades ago, in the search for convenient stable products through removal of water, resulted in the so-called modern intermediate moisture foods. These foods rely heavily on the addition of humectants and preservatives to prevent or reduce the growth of microorganisms. Since then, this category of products has been subjected to continuous revision and discussion.

Definitions of IMF in terms of a_w values and moisture content vary within wide limits (0.6-0.90 a_w, 10-50% moisture), and the addition of preservatives provides the margin of safety against spoilage organisms tolerant to low a_w. Of the food poisoning bacteria, *Staphylococcus aureus* is one of the organisms of high concern since it has been reported to tolerate a_w as low as 0.83-0.86 under aerobic conditions. Many of the considerations on the significance of microorganisms in IMF are made in terms of a_w limits for growth. However, microbial control in IMF does not only depend on a_w but on pH, *E*h, *F* and *T* values preservatives, competitive microflora, etc., which also exert an important effect on colonizing flora.

Fruits Preserved Under IMF Concept

The application of IMF technology has been very successful in preserving fruits and vegetables without refrigeration in most Latin American countries. For instance, the addition of high amounts of sugar to fruits during processing will create a protective layer against microbial contamination after the heat process. The sugar acts as a water activity depressor limiting the capability of bacteria to grow in food. IMF foods are those with a_w in the range of 0.65 to 0.90 and moisture content between 15% and 40%. Food products formulated under this concept are stable at room temperature without thermal processing and can be generally eaten without rehydration. Some processed fruits and vegetables are considered IMF foods. These include cabbage, carrots, horseradish, potatoes, strawberries, etc.; their water activities at 30°C follow:

Foods	a_w
Cabbage	0.64
	0.75
Carrots	0.64
	0.75
Horseradish	0.75
Potatoes	0.75
	0.64
Strawberries	0.65
	0.75

Under these conditions, bacterial growth is inhibited but some moulds and yeast may grow at a_w greater than 0.70. In addition, chemical preservatives are generally used to inhibit the growth of moulds and yeasts in fruits and vegetables.

Advantages and disadvantages of IMF preservation

Advantages

\Intermediate moisture foods have an a_w range of 0.65-0.90, and thus water activity is their primary hurdle to achieving microbial stability and safety. IMF foods are easy to prepare and store without refrigeration. They are energy efficient and relatively cheap. They are not readily subject to spoilage, even if packages have been damaged prior to opening, as with thermostabilized foods, because of low a_w. This is a plus for many developing countries, especially those in tropical climates with inadequate infrastructure for processing and storage, and offers marketing advantages for consumers all over the world.

Disadvantages

\Some IMF foods contain high levels of additives (i.e., nitrites sulphites, humectants, etc.) that may cause health concerns and possible legal problems. High sugar content is also a concern because of the high calorific intake. Therefore, efforts are been made to improve the quality of such foods by decreasing sugar and salt addition, as well as by increasing the moisture content and a_w, but without sacrificing the microbial stability and safety of products if stored without refrigeration. This may be achieved by an intelligent application of hurdles.

Fruit products from intermediate moisture foods (IMF) appear to have potential markets. However, application of this technology to produce stable products at ambient temperature is limited by the high concentration of solutes required to reduce water activities to safe levels. This usually affects the sensory properties of the food.

Combined Methods for Preservation of Fruits and Vegetables

Food preserved by combined methods (hurdles) remains stable and safe even without refrigeration, and is high in sensory and nutritive value due to the gentle process applied. Hurdle technology is the term often applied when foods are preserved by a combination of processes. The hurdle includes temperature, water activity, redox potential, modified atmosphere, preservatives, etc. The concept is that for a given food the bacteria should not be able to "jump over" all of the hurdles present, and so should be inhibited. If several hurdles are used simultaneously, a gentle preservation could be applied, which nevertheless secures stable and safe foods of high sensory and nutritional properties. This is because different hurdles in a food often have a synergistic (enhancing) or additive effect. For instance, modified foods may be designed to require no refrigeration and thus save energy. On the other hand, preservatives (e.g., nitrite in meats) could be partially replaced by certain hurdles (such as water activity) in a food. Moreover, a hurdle could be used without affecting the integrity of food pieces (e.g., fruits) or in the application of high pressure for the preservation of other foods (e.g., juices). Hurdle technology is applicable both in large and small food industries. In general, hurdle technology is now widely used for food design in making new products according to the needs of processors and consumers. For instance, if energy preservation is the goal, then energy consumption hurdles such as refrigeration can be replaced by hurdles (a_w, pH, or E_h) that do not require energy and still ensure a stable and safe product.

The hurdle effect is an illustration of the fact that in most foods several factors (hurdles) contribute to stability and safety. This hurdle effect is of fundamental importance for the preservation of food, since the hurdles in a stable product control microbial spoilage and food poisoning as well as undesirable fermentation.

General Description of Combined Methods

Increasing consumer demand for fresh quality products is turning processors to the so-called minimally processed products (MP), an attempt to combine freshness with convenience to the point that even the traditional whole, fresh fruit or vegetable is being packaged and marketed in ways formerly reserved for processed products. According to these authors, the widely accepted concept of MP refrigerated fruits involves the idea of living respiring tissues. Because MP refrigerated products can be raw, the cells of the vegetative tissue may be alive and respiring (as in fruits and vegetables), and biochemical reactions can take place that lead to rapid senescence and/or quality changes. In these products, the primary spoilage mechanisms are microbial growth and physiological and biochemical changes, and in most cases, minimally processed foods are more perishable than the unprocessed raw materials from which they are made.

The technology for shelf-stable high moisture fruit products (HMFP) is based on a combination of inhibiting factors to combat the deleterious effects of microorganisms in fruits, including additional factors to reduce major quality losses from reactions. In order to select a combination of factors and levels, the type of microorganism and quality loss from reactions that might occur must be anticipated. Minimal processing may encompass pre-cut refrigerated fruits, peeled refrigerated whole fruits, sous vide dishes, which may include pre-heated vegetables and fruits, cloudy and clarified refrigerated juices, freshly squeezed juices, etc. All of these products have special packaging requirements coupled with refrigeration. These products, apart from special handling, preparation, and size reduction operations, might also require special distribution and utilization operations such as Controlled atmosphere/Modified atmosphere/air flow rate/vacuum storage (O_2, CO_2, N_2, CO, C_2H_2, H_2O controls), computer controlled warehousing, retailing and food service, communications network, etc. HMFP fruits are less sophisticated than MPR fruits and should be priced lower when introduced commercially. Careful selection of these processes should of course be made to find the appropriate methods suited to a particular rural or village situation.

Recommended Substances to Reduce a_w in Fruits

Glucose

Glucose is not a very good humectant due to the lower water holding capacity (WHC), which makes it difficult to obtain the isotherm curve at low a_w.

Fructose

Fructose has a higher water activity reduction capacity and therefore is more desirable as a humectant in stabilizing food products.

Sucrose

Sucrose is one of the most studied sugars and is widely used in food systems, in the confectionary industry, both in the U.S. and Europe, but has a lower water activity reduction capacity compared to fructose.

Other humectants

Based solely on the water activity reduction capacity, sorbitol and fructose are the most desirable humectants. Sucrose has the third best reduction capacity and lactose the poorest. The amorphous form absorbs more water at specific a_w than the corresponding crystalline form. NaCl and KCl salts appear to be superior humectants at a high range of a_w. The increased a_w lowering ability exhibited by the salts may be explained by the smaller molecular weight, increasing the ability to bind or structure more water.

Other sugars used as humectants in food stability include lactose and sorbitol. The amorphous form absorbs more water at specific a_w than the crystalline form. Polyols are better humectants than sugars because of their greater water activity reduction capacity and are less hygroscopic than sugars. The most widely used polyols as humectants in foods are 1,3- butyleneglycol, propylene glycol, glycerol, and polyethylene glycol 400.

Recommended substances to reduce pH

Organic acids

Organic acids, whether naturally present in foods due to fermentation or intentionally added during processing, have been used for many years in food preservation. Some organic acids behave primarily as fungicides or fungistats, while others tend to be more effective at inhibiting bacterial growth. The mode of action of organic acids is related to the pH reduction of the substrate, acidification of internal components of cell membranes by ionization of the undissociated acid molecule, or disruption of substrate transport by alteration of cell membrane permeability. The undissociated portion of the acid molecule is primarily responsible for antimicrobial activity; therefore, effectiveness depends upon the dissociation constants (pKa) of the acid. Organic acids are generally more effective at low pH and high dissociation constants. The most commonly used organic acids in food preservation include: citric, succinic, malic, tartaric, benzoic, lactic, and propionic acids.

Citric acid is present in citrus fruits. It has been demonstrated that citric acid is more effective than acetic and lactic acids for inhibiting growth of thermophilic bacteria. Also, combinations of citric and ascorbic acids inhibit growth and toxin production *of C. botulinum* type B in vacuum-packed cooked potatoes.

Malic acid is widely found in fruits and vegetables. It inhibits the growth of yeasts and some bacteria due to a decrease in pH.

Tartaric acid is present in fruits such as grapes and pineapples. The antimicrobial activity of this acid is attributed to pH reduction.

Benzoic acid is the oldest and most commonly used preservative. It occurs naturally in cranberries, raspberries, plums, prunes, cinnamon, and cloves. As an additive, sodium salt in benzoic acid is suitable for foods and beverages with pH below 4.5. Benzoic acid is primarily used as an antifungal agent in fruit-based and fruit beverages, fruit products, bakery products, and margarine.

Lactic acid is not naturally present in foods; it is formed during fermentation of foods such as sauerkraut, pickles, olives, and some meats and cheeses by lactic acid bacteria. It has been reported that lactic acid inhibits the growth of spore forming bacteria at pH 5.0 but does not affect the growth of yeast and moulds.

Propionic acid occurs in foods by natural processing. It is found in Swiss cheese at concentrations up to 1%, produced by *Propionicbacterium shermanii*. The antimicrobial activity of propionic acid is primarily against moulds and bacteria.

norganic acids

Inorganic acids include hydrochloric, sulphuric, and phosphoric, the latter being the principal acid used in fruit and vegetable processing). They are mainly used as buffering agents, neutralizers, and cleaners.

Fermentation by-products

Fermentation by-products are formed during fermentation of fruits and vegetables, as in sauerkraut processing, pickling, and wine making. One by-product, lactic acid, is formed during fermentation of cabbage or cucumbers. This acid decreases the pH of fruits and vegetables, producing the characteristic flavour of sauerkraut, and acts as a controller of pathogens that may develop in the final fermented product.

Recommended chemicals to prevent browning

Sulphites, bisulphites, and metabisulphites

Sodium bisulphite is a potential browning inhibitor in fruit and vegetable products (e.g., peeled potatoes and apples). This preservative when used in food production can delay or prevent undesirable changes in the colour, flavour, and texture of fresh fruits and vegetables, potatoes, drinks, wine, etc. Potassium bisulphite is used in a similar way to sodium bisulphite, and is used in the food industry to prevent browning reactions in fruit and vegetable products.

Sulphites, bisulphites, and metabisulphites of both sodium and potassium together with gaseous sulphur dioxides are all chemically equivalent. Sulphite levels in processed foods are expressed as SO_2 equivalents, and range from zero to about 3000 ppm in dry weight. Dehydrated, light coloured fruits (e.g., apples, apricots, bleached raisins, pears, and peaches) contain the greatest amounts in this range. Dehydrated vegetables and prepared soup mixes range from a few hundred to about 2000 ppm; instant potatoes contain approximately 400 ppm. The dose for wine is about 100-400 ppm and for beer about 2-8 ppm. The maximum legal sulphite level in wines permitted by the Food and Drug Administration (FDA) is 300 ppm. In the U.S. most wines have a sulphite level of 100 ppm.

Sulphites are highly effective in controlling browning in fruits and vegetables, but are subject to regulatory restrictions because of adverse effects on health. Sulphites inhibit non-enzymatic browning by reacting with carbonyl intermediates, thereby preventing further reaction. Sulphite levels in foods vary widely depending on the application. Residual levels never exceed several hundred per million but could reach 100 ppm in some fruits and vegetables.

The maximum sulphur dioxide levels in fruit juices, dehydrated potatoes, and dried fruits permitted by the FDA are 300, 500, and 2000 ppm, respectively.

Recommended additives to inhibit microorganisms

Potassium sorbate

Potassium sorbate is a white crystalline powder that has greater solubility in water than sorbic acid, which may be used accordingly in making concentrates for dipping, spraying, or metering fruit and vegetable products. It has antimycotic actions similar to sorbic acid, but usually 25% more potassium sorbate must be used than sorbic acid to secure the same protection.

The common salt of potassium sorbate was developed because of its high solubility in water, which is 58.2% at 20°C. In water, the salt hydrolysis yielded is the active form. Stock solutions of potassium sorbate in water can be concentrated up to 50%, which can be mixed with liquid food products or diluted dips and sprays. Sorbates are effective in retarding the growth of many food spoilage organisms. Sorbates have many uses because of their milder taste, greater effectiveness, and broader pH range (up to 6.5), when compared to either benzoate or propianate. Thus, in foods with very low pH, sorbate levels as low as 200 ppm may give more than adequate protection. The solubility of potassium sorbate is 139 g/100 mL at 20°C; it can be applied in beverages, syrups, fruit juices, wines, jellies, jams, salads, pickles, etc.

Sodium benzoate

The use of sodium benzoate as a food preservative has been limited to products that are acid in nature. Therefore, it is mainly used as an antimycotic agent (most yeasts and moulds are inhibited by 0.05-0.1%). The benzoates and parabenzoates have been used primarily in fruit juices, chocolate syrup, candied fruit peel, pie fillings, pickled vegetables, relishes, horseradish, and cheeses. Sodium benzoate is more effective in food systems where the pH is as low as 4.0 or below.

Other additives

Other naturally antimicrobial compounds found in fruits and vegetables include:

Vanillin (4-hydroxi-3-methoxybenzaldehyde) is found primarily in vanilla beans and in the fruit of orchids *(Vanilla planifola*, *Vanilla pompona*, or *Vanilla tahitensis*). Vanillin is most active against moulds and non-lactic acid gram-positive bacteria. The effectiveness of vanillin against certain moulds such as *A. flavus*, *A. niger*, *A. ochraceus*, *or A. parasiticus* has been demonstrated in laboratory media, as well as its effectiveness against yeasts such as*Saccharomyces cerevisiae, Pichia membranaefaciens, Zygosaccharomyces bailii, Z. rouxii* and *Debaryomyces hansenii.*

Allicin is an antimicrobial present in the juice vapour of garlic. This compound is effective in inhibiting the growth of certain pathogenic bacteria such as *B. cereus*, C. botulinum, *E. coli*, *Salmonellae*, *Shigellae*, *S. aureus*, *A. flavus*,*Rhodotorula*, and *Saccharomyces*.

Cinnamon and eugenol are reported to have an inhibitory effect on the spores of *Bacillus anthracis*. Also, cinnamon was found to inhibit the growth of the aflatoxin of *A. parasiticus.* Aqueous clove infusions of 0.1 to 1.0% and 0.06% eugenol were reported to inhibit the growth of germinated spores of *B. subtilis* in nutrient agar.

*Oregano, thyme, and rosemary have been found to have inhibitory activity against certain bacteria an*d moulds due to the presence of antimicrobial compounds in their essential oils (e.g., terpenes, carvacol, and thymol).

Recommended thermal treatment for food preservation

The role of heat

The main function of heat in food processing is to inactivate pathogenic and spoilage organisms, as well as enzyme inactivation to preserve foods and extend shelf life. Other advantages of heat processing include the destruction of anti-nutritional components of foods (e.g., trypsin inhibitors in legumes), improving the digestibility of proteins, gelatinization of starches, and the release of niacin. Higher temperatures for shorter periods achieved the same shelf life extension as food treated at lower temperatures and longer periods, and allowed retention of sensory and nutritional properties.

Hot water

Hot water plays an important role in the sanitation of food products before processing. Some food products are treated with hot water to eliminate insects, and to inactivate microorganisms and enzymes. Foods are retained in a water blancher at 70-100°C for a specific time and then removed to a dewatering and cooling system.

Steam

Steam is a more effective means than hot water for blanching foods such as fruits and vegetables. This method is especially suitable for foods with large areas of cut surfaces. It retains more soluble compounds and requires smaller volumes of waste for disposal than those from water blanchers. This is particularly so if air-, rather than water-cooling is used. Furthermore, steam blanchers are easier to clean and sterilize.

Effects of heat on aerobic and anaerobic mesophylic bacteria, yeasts, and moulds

Temperatures ranging from 10 to 15°C above the optimum temperature for growth will destroy vegetative cells of bacteria, yeasts, and moulds. Most vegetative cells, as well as viruses, are destroyed when subjected to temperatures of 60 to 80°C for an appropriate time. Somewhat higher temperatures may be needed for thermophilic or thermoduric microorganisms. All vegetative cells are killed in 10 min at 100°C and many spores are destroyed in 30 min at 100°C. Some spores, however, will resist heating at 100°C for several hours.

High Moisture Products

The importance of considering the combined action of decreased water activity with other preservation factors as a way to develop new improved foodstuffs has been studied. Leistner introduced the hurdle concept, or hurdle effect, to illustrate the fact that in most foods, a combination of preservation parameters (hurdles) accounts for their final microbial stability and safety. Since then, these concepts have been improved to the point that depending on the acting hurdles of high relevance to a particular product, shelf-stability can be accomplished by a careful handling of complementary hurdles. For instance, the pH of IMF should be as low as palatability permits, and whenever possible, below pH 5.0. Undoubtedly, this imposes a limitation not only on colonizing microflora, but also on foodstuffs, since pH cannot be reduced in many products without flavour impairment. Even at low pH values and low a_w, certain yeast and mould species that can tolerate high solute concentrations might pose a risk to the stability of IMF.

Fruits are a good example of foodstuffs that accept pH reduction without affecting the flavour significantly. Important developments on IMF based on fruits and vegetables are reported elsewhere. The extensive research conducted in India by Dr. Jayaraman and co-workers has generated important information on this product category. Technological problems have prevented IMF from further development. Also, consumer health concerns associated with the high levels of humectants and preservatives used, have contributed to this situation. This last issue has become more important in recent years due to greater public awareness of food safety concerns. Additionally, consumers are searching for fresh-like characteristics in products. The food industry has responded to these demands with the so-called minimally processed fruits and vegetables, which have become a widespread industry. Consequently, safety considerations are being addressed seriously by food microbiologists.

Different approaches can be explored for obtaining shelf-stability and fresh-likeness in fruit products. Commercial, minimally processed fruits are fresh (with high moisture), and are prepared for convenient consumption and distribution to the consumer in a fresh-like state. Minimum processing includes preparation procedures such as washing, peeling, cutting, packing, etc., after which the fruit product is usually placed in refrigerated storage where its stability varies depending on the type of product, processing, and storage conditions. However, product stability without refrigeration is an important issue not only in developing countries but in industrialized countries as well. The principle used by Leistner for shelf-stable high moisture meats (a_w >0.90), where only mild heat treatment is used and the product still exhibits a long shelf life without refrigeration, can be applied to other foodstuffs. Fruits would be a good choice. Leistner states that for industrialized countries, production of shelf-stable products (SSP) is more attractive than IMF because the required a_w for SSP is not as low and less humectants and/or less drying of the product is necessary.

If fresh-like fruit is the goal, dehydration should not be used in processing. Reduction of a_w by addition of humectants should be employed at a minimum level to maintain the product in a high moisture state. To compensate for the high moisture left in the product (in terms of stability), a controlled blanching can be applied without affecting the sensory and nutritional properties; pH reductions can be made that will not impair flavour; and preservatives can be added to alleviate the risk of spoilage by microflora. In conjunction with the above mentioned factors, a slight thermal treatment, pH reduction, slight aw reduction and the addition of antimicrobials (sorbic or benzoic acid, sulphite), all placed in context with the hurdle principle applied to fruits, make up an interesting alternative to IMF preservation of fruits, as well as to commercial minimally processed fruits.

Alzamora et al. conducted pioneer work aimed at obtaining shelf-stable peaches and pineapple. Considerable research has been made within the CYTED Program and the Multinational Project on Biotechnology and Food of the Organization of American States (OAS) in the area of combined methods geared to the development of shelf-stable high moisture fruit products.

Over the last decade, use of this approach has led to important developments of innovative technologies for obtaining shelf-stable "high moisture fruit products" (HMFP) storable for 3-8 months without refrigeration. These new technologies are based on a combination of inhibiting factors to combat the deleterious effects of microorganisms in fruits, including additional factors to diminish major quality loss in reactions rates. Slight reduction of water activity (a_w 0.94-0.98), control of pH (pH 3.0-4.1), mild heat treatment, addition of preservatives (concentrations £ 1,500 ppm), and antibrowning additives were the factors selected to formulate the preservation procedure. These techniques were preceded by the pioneer work of Leistner on the combined effects of several factors applied to meat products - named "hurdle" technology. Microbiological preservation with these combined techniques, by gently applying individual stress factors to control microbial growth, avoid the severity of techniques based on the employment of only one conservation factor.

Preliminary Operations

Preliminary operations involve washing, selecting, peeling, slicing, and general blanching of fresh fruits. Fresh produce must be processed between 4 and 48 hours after harvest to prevent the growth of spoilage microorganisms.

Washing: This operation involves eliminating dirt from the material before it passes through the processing line. Fruits are washed with potable water by immersion, spraying or brushing to eliminate the soil. Sodium hypochlorite is usually added to the water at a rate of 10% (v/v). The effectiveness of chlorine is enhanced by using a low pH, high temperature, pure water, and the correct contact time.

Fruit selection: The cleaned product is selected for processing by separating the damaged fruits from those free of defects and disease. The fruit must be of a uniform size, form, colour, and maturity.

Peeling: This operation consists of removing the skin from the fruit (usually by hand) using a sharp knife. There are several peeling methods available, but on an industrial scale, peeling is normally accomplished mechanically (e.g., rotating carborundum drums) and chemically, or with high-pressure steam peelers.

Slicing: This operation involves cutting the fruit into several uniform pieces, which is more convenient than handling the entire fruit. This is accomplished manually with a sharp knife or with special cutting machines that produce clean, neat slices.

Blanching: This is a critical control operation in the processing of high moisture fruit products (HMFP). It is an early step for processing of several fruits. Destruction of contaminating organisms is not the treatment's main objective, but it occurs nevertheless because the temperature used is lethal to yeast, most moulds, and aerobic natural flora. Many microorganisms can survive heat treatment but are sensitive to other hurdles like pH and water activity (a_w). A 60 to 99% reduction in the microbial load of HMFP for papaya, pineapple, strawberry, and mango has been reported. For mangoes, the microbial counts decreased from 14.3×10^3 cfu/g in the fresh fruit to 1.3×10^3 cfu/g after blanching. The blanching temperatures were between 85 and 100°C for very short periods, usually 3 to 5 minutes.

Desired a_w and Syrup Formulation

The desired a_w is determined by equilibrium of the components in the food system. This includes the addition of water, sugar (sucrose, glucose, or fructose), and chemicals such as citric acid, sodium bisulphite, and potassium sorbate, etc. The levels of sodium bisulphite and potassium sorbate in the system can be used at 150 and 1000 ppm, respectively. Once the system is in equilibrium, the a_w can be measured using an automatic water activity meter to an accuracy of + or - 0.005. These instruments are now available over specified ranges as laboratory or portable hand meters.

Calculus required

To determine the desired a_w in syrup (a_w equilibrium), the Ross equation is used:

$$a_{w\ equilibrium} = (a°w)fruit \cdot (a°w)\ sugar \quad (1)$$

where $a°_{w\ fruit}$ is the water activity of the fruit and $a°_{w\ sugar}$ is the water activity of sugar, both calculated at the total molality of the system. The product of the molality of sucrose in the fruit water and solution must equal the desired water activity in equilibrium. The $a°_W$ values of the sugar are obtained using the Norrish equation:

$$a°_{w\ sucrose} = X_1 \exp(-kX_2^2) \quad (2)$$

where k is a constant for sugars, X_1 and X_2 are the molar fractions of water and sugar, respectively. Some K values for common sugars and polyols are listed in Table 1.

Table 1. Values of Norrish constant for common sugars and polyols.

Sugars	*k*
Sucrose	6.47 ±0.06
Maltose	4.54 ±0.02
Glucose	2.25 ± 0.04
L actose	10.2
Polyols	
Sorbitol	1.65 ±0.14
Glycerol	1.16 ±0.01
Mannitol	0.91 ±0.27
Propylene Glycol	4.04
Arabitol	1.41

Phosphoric or citric acids are generally used to reduce the syrup's pH so that the final pH of the fruit-syrup system is in equilibrium in the desired range (3.0 to 4.1). Monitoring of a_w and pH in the fruit and syrup until constant values for these parameters are reached can determine the time to equilibrate the system. This may be from three to five days at constant room temperature depending on the size of fruit pieces.

Packaging Methods for Minimally Processed Products

The purpose of food packaging is to maintain quality and to obtain shelf life extension of products by reducing mechanical damage and retarding microbial spoilage. Three types of packaging methods exist for minimally processed products: unit packaging, transport packaging, and loading packaging. Other packaging methods are vacuum and modified atmospheres.

Packaging with Small Units

This type of packaging method uses (1) closed plastic bags, (2) rigid or semi-rigid plastic trays zipped in upper part with polymeric plastic film, (3) covered trays for distribution of products to institutions (e.g., hotels, restaurants, and food shops) and small business consumer markets, (4) perforated or unperforated PE or PVC bags, (5) shallow trays, (6) cartons, and (7) thermo-formed plastic tubs or expanded PS containers covered/sealed with polymeric film.

Two of the main requirements for this type of packaging are its permeability characteristics to any gases present and to water vapour. Other important considerations include: appearance (brightness and transparency), texture, resistance to water permeability, resistance to impact and deformation, thermo-seal capacity, ease in forming/fabricating/filling, and utilization of production equipment. Plastic containers are also light weight, sometimes reusable, tough, hygienic, and rigid containers can be stacked.

Transport the Package

Packaging for transport of products is dominated by sealed cartons made from corrugated paper. These types of packages provide good resistance to mechanical damage of fresh fruit, and facilitate manual handling of fresh fruits during transportation to markets. The cartons are made of paper, > 0.2 mm thick, obtained from vegetable cellulose bonded either in three layers with the middle one corrugated or in five layers with the second and fourth layers corrugated. Both systems provide a strong and rigid material.

Loading Packaging Units

This type of packaging implies the use of palletization of packages to reduce the cost of handling. In this way, the mechanical work of loading and unloading by carriers is facilitated, permitting better utilization of storage space and reducing mechanical damage during transportation.

Vacuum and Modified Atmosphere Packaging

Vacuum packaging of fresh commodities involves eliminating (at least some) the air in the package using a suction machine. This method reduces the level of both oxygen and nitrogen in the package, prolonging the shelf life of fruits for extended periods.

Vacuum packaging is used in modified atmosphere packaging (MAP) of fruits and vegetables. The basic principle behind modified atmosphere packaging (MAP) is that a modified atmosphere can be created passively by correctly using permeable packaging materials, or actively, by using a special gas mixture combined with such materials. The purpose of both is to create an optimal gas balance inside the package, where the respiration activity of a product is as low as possible; on the other hand, the oxygen concentration and carbon dioxide levels are not detrimental to the product. In general, the aim is to have a gas composition of 2-5% CO_2, 2-5% O_2, and the rest nitrogen. A problem that arises when using MAP is the restricted availability of permeable material in the market, as only a few materials are permeable enough to match the respiration of fruits and vegetables. Most films do not result in optimal O_2 and CO_2 atmospheres, especially when the product has high respiration.

However, one solution is to make microholes of defined sizes and defined quantity in the material to avoid anaerobiosis. Other solutions are to combine ethylene vinyl acetate with orientated polypropylene and low-density polyethylene, or to combine ceramic material with polyethylene. Both composite materials have significantly higher gas permeability than polyethylene or orientated polypropylene. They are used a lot in the packaging of salads, although gas permeability should be higher.

One interesting MAP method is called moderate vacuum packaging (MVP). In this system, respiring produce is packed into a rigid, airtight container at less than 0.4 of normal atmospheric pressure (40 kPa) and stored at refrigerated temperatures (4-7°C). The initial gas composition is that of normal air (21% O_2, 0.04 CO_2, and 78% N_2) but is at reduced partial gas pressure. The lower O_2 availability stabilizes the produce quality by slowing its metabolism and the growth of microorganisms.

Storage, and Use of Fruits Preserved by Combined Methods

Open vs. Refrigerated Vehicles

Open vehicles are mainly used to transport fresh produce over short distances from the field to packinghouses, retail markets, or the processing plant directly. The fruit must be protected against mechanical damage and sunlight. Therefore, the produce should be transported at night or in the early morning. Refrigerated vehicles should to be used to transport fruits. In this case, the vehicle must be equipped with an efficient cooling system, adequate distribution and circulation of air, relative humidity and temperature sensors, and it must be well insulated

Unloading

Unloading of fruits from vehicles can be done by hand or mechanical means. Forklifts are used to unload vehicles in which packages of fruits have been palletized. During unloading, care must be taken in handling the packages to avoid dropping, which can cause damage to the package and bruising of the fruit upon impact. Impact injury may not be visible on the surface; so careful control is needed to prevent its occurrence.

Storage Temperature vs. Shelf Life

Refrigeration is the largest hurdle for MPF and the most difficult to control. During transport, handling, and storage of fruits by consumers, temperature is often not adequately maintained, resulting in spoilage. Food products exposed to elevated temperatures where refrigeration is the only factor of preservation are more susceptible to damage and spoilage, and thus the shelf life is very short. Optimum refrigeration temperatures for fruits and vegetables vary widely. Some authors suggest between 10 and 15°C for cooling and between 2 and 5°C for refrigeration. Table 2 exhibits the optimum temperatures for storage of refrigerated fruits. Data is given for fresh products but the temperatures could change according to the process applied to a particular fruit. The data for recommended shelf life and the safety of minimally processed refrigerated fruits (MPRF) is still not available for public use. In general, MPRF products are classified as food products with prolonged shelf life where refrigeration is the preservation method most commonly used for this purpose. The stability of fruits without refrigeration is an important issue in developing and industrialized countries. Minimally processed refrigerated fruits (MPRF) are not shelf stable at ambient temperatures and should be distributed and marketed in a reliable cold chain for safety and retention of sensory and nutritional quality. Hurdle technology has proved effective in preserving tropical and sub-tropical fruits with fresh-like properties. This technique includes blanching as an MP preservation method and excludes the use of refrigeration.

Table 2. Optimum temperatures for storage of refrigerated fruits.

Fruit	*Temperature*
Apples	30-31°F (-1.1 to - 0.6°C)
Varieties sensitive to refrigeration	38-40°F (3.3. to 4.4 °C)
Apricots	31-32°C (-0.6 to 0°C)
Green plantains	56-58°F (13.3 to 14.4 °C)
Bush berries, blueberries, strawberries	32°C (0 °C)
Cherries	30-32°F (1.1 to 0 °C)
Grapefruits	58-60°F (14.4 to 15.6 °C)
Lemons	58-60°F (14. 4 to 15.6°C)
Limes	45-50°F (7.2 to 10 °C)
Oranges	38-44°F (3.3 to 6.7 °C)
Tangerines	32°F (0 °C)
Coconuts	32-35°F (-0.6 to 0°C)
Dates	32°F (0 °C)
Figs	31-32°F (-0.6 to 0°C)
Grapes	30-31°F (-1.1 to 0.6°C)
Mangoes	55°F (12.8 °C)
Honeydew	45-50°F (7.2 -t o 10°C)
Cantaloupe	32-40°F (0 -t o 4.4°C)
Watermelon	40-32°F (-0.6 to 0°C)
Papayas	31°F-32°F (-0.6 to 0°C)
Peaches and nectarines	32°F (0 °C)
Pears	29-31°F (-1.7 to -0.6°C)
Pineapples	45-47°F (7.6 to 8.3 °C)
Prunes	31°F-32°F (-0.6 to 0°C)
Grenade	32°F (0 °C)
Quince fruit	32°F (0 °C)

The shelf life of high moisture fruits or purées is extended from at least 3 months to 8 months at room temperature. These fruit products are quite different from intermediate moisture fruits (high sugar candied fruits) because of a lower sugar concentration (24-28% w/w vs. »(70% w/w red. sugars) and higher moisture content (55-77% w/w vs. 20-40% w/w) that resembles canned fruit. They can be eaten as received or used as bulk for out-of-season processing, in confectionery, bakery goods, and dairy products, or for preserves, jams, and jellies.

Repackaging considerations

Minimally processed fruit products can be repackaged from bulk containers into small packages such as glass or plastic jars, and high-density polyethylene bags for retail markets and consumer distribution. The stabilized fruit products can be processed in the form of slices, chunks, whole fruit, marmalades, or nectars.

Syrup reconstitution and utilization

Syrup reconstitution is needed for repackaging of MPFP, which requires the addition of sugar, and additives to adjust the water activity, pH, and control of browning reaction. The syrup covers the fruit inside the package and protects against microbial contamination. It should have a pH between 3.0 and 4.1. The tank holding the fruit and syrup prior to repackaging should be maintained at constant room temperature for 3 to 5 days during equilibration.

Optimal utilization of the final product

The final MPF product can be eaten as received or used in bulk for off-season processing, in confectionery, bakery goods, and dairy products, or for preserves, jams, and jellies. Fruit pieces can be utilized for salads, barbecue sauces, pizzas, fruit drink formulations, etc.

Quality Control

Recommended Microbiological Tests

Several microbiological tests should be implemented in the processing area according to the Good Manufacturing Practice (GMP). Microbiological tests also apply to working personnel who manipulate and prepare the fruit products.

Total aerobic counts (TAC): TAC is performed in petri dishes with standard plate count agar (SPCA). These are plated with a spread from the hair, fingerprints, shoe soles, work tables, utensils, and skin of workers with the aid of a wet cue tip, which has been impregnated with a sterile peptone solution (1% v/v). The impregnated cue tip is passed through the desired area being controlled, then spread onto the agar surface in the petri dish. The plates are incubated at 35-37°C ± 2°C for 18 to 24 hours.

Mould and Yeast counts (MYC): To count mould and yeast cells, plates with potato-dextrose agar are plated with the same infected areas described above and incubated for 5-7 days at 25-30°C ± 2°C.

Knowledge of the combined effect of the preservation factors used for high moisture fruit products (HMFP) on the growth and survival of certain key microorganisms that may pose risks to the quality and safety of HMFP is of great interest in the design of this technology.

Nutritional Changes

Very small changes in the nutritional characteristics of MPF are experienced during processing and storage, due to the mild heat treatment applied. Blanching does not affect the nutritional properties, but it does inactivate the enzymes and provide some reduction of indigenous flora.

Changes in Sensory Attributes and Acceptability

Changes in flavour, texture, odour, and colour have not been reported in high moisture minimally processed fruit products (HMPFP), such as papaya, peach, pineapple, and mango. In general, the average scores presented in Table 4 correspond to products that have good acceptability.

Table 4 Sensory characteristics of shell life stable high moisture papaya, peach, pineapple, and mango.

Attribute	*Average score*
Flavour	6.65-7.70
Odour	5.80-6.80
Texture	6.70-8.07
Colour	6.46-7.10
Overall impression	6.73-7.63

Texture received the highest scores followed by flavour, colour, and general impression, indicating that combined method technology is a viable alternative in fruit preservation. These parameters are usually judged by using a small trained panel or a larger group of non-trained volunteers. A numerical scale is given for each attribute and the response of each judge is recorded. A scorecard is prepared with a hedonic scale ranging from 0 to 9 points, which is presented to each judge. Nine is the highest score, "like very much", and zero (0) is the lowest score, "dislike very much". Samples are identified with a code number selected at random.

References

Arthey, D. and Ashurst, P.R. 1996. *Fruit Processing*. Blackie Academic & Professional, London, 142-125.

FAO. 1988. *Packaging for Fruits, Vegetables and Root Crops*. FAO. Rome.

Nagy, S. and Shaw, P.E.. *Tropical and Subtropical Fruits. Composition, Properties and uses*. AVI Publishing, Westport, Connecticut, 127.

Ryall, A.L. and Pentzer, W.T. 1979. *Handling, Transportation and Storage of Fruits and Vegetables*, 320-332.

Sofos, J.N. 1989. *Sorbate Food Preservatives*. CRC Press, Boca Raton, 16-17.

Thompson, A.K. 1996. *Post-harvest Technology of Fruit and Vegetables*. Blackwell Science, Berlin.

3

Methods of Meat Preservation

Although the demand in developing countries for animal proteins is increasing, animal production has failed to keep pace with the growth in demand and to make full use of its potential in developing countries.

The nutrition of resource-poor rural producers can be improved both directly by consumption of animal food products or indirectly by enabling the purchase of food with returns from animal product sales.

The average per caput supply of meat in developing countries is very low. In 1979 and 1986 it amounted to 12.7 and 15.4 kg/caput/year respectively as compared to the developed countries with 74.3 and 77.9 kg/ caput/ year respectively. The average per caput supply in developing countries in Africa reached the average level of all developing countries in 1979 with 12.9 kg/caput/ year but in 1986 remained at 11.8 kg/caput/year, below the level of the developing countries combined and even below the 1979 level.

While efforts are increasing to support animal production in developing countries, they are not matched by similar efforts to use preservation to overcome seasonal variation in meat supply. In addition the existing conditions for slaughtering and meat handling in rural areas which cause quality deterioration and post-harvest losses of meat- and food-borne diseases in consumers must be improved.

In fact there is a lack of effort to provide knowledge and skills in adequate hygienic slaughtering, meat cutting and handling under rural conditions. Taking into account that an uninterrupted cold chain for meat cannot be expected in many developing countries in the near future, the absence of meat preservation techniques presents a serious constraint to the development of viable meat production by resource-poor rural livestock producers.

Adequate meat preservation complements a marketing system which by necessity has been adapted to a fast throughput of fresh meat and which does not facilitate the use of surplus meat in periods of meat shortage.

While this publication is mainly intended to disseminate information on traditional methods of meat preservation in Africa for teachers and instructors, it also addresses aspects of hygienic slaughtering under rural conditions. Reference is also made to FAO's work on small-scale slaughterhouses, raw materials for preserved meat, principles of meat preservation by thermal treatment, packaging methods and basic methods of quality control.

Slaughtering and Raw Materials for Meat Preservation

Product quality and shelf-life of preserved meat and meat products depend on the microbiological and biochemical status of the carcass meat used for processing. In other words, the raw material must be as clean as possible and derived from slaughter animals in good and healthy condition.

Obviously the way of slaughtering animals plays an important role. Many irreversible quality losses, especially with regard to the hygienic quality, originate from improper slaughtering and carcass handling. Slaughter techniques, particularly under rural conditions, cannot and need not be sophisticated, but they must allow carcass meat which complies with basic hygienic requirements to be obtained whether slaughtering at the farm, or in slaughterslabs or slaughterhouses.

Careless slaughter and meat handling result in:

- improper and insufficient bleeding of the animals, leaving a relatively high amount of blood in the muscle. As a consequence the meat does not reach the necessary degree of acidity and its shelf-life is reduced;
- improper dehiding of the carcass, leading to heavy contamination of the meat surfaces by frequent contact with the personnel, polluted floors and dirty tools and equipment;

- improper evisceration through accidental opening of stomachs and tripes, leading to contamination spreading to internal and external surfaces of the carcass;
- contamination of the meat when carcasses are split on the ground;
- contamination of the meat during transport, when carcasses or parts of them make contact with unclean clothes, hooks, bars or containers;
- contamination of the meat stored under unhygienic conditions and because of faulty handling in meat markets, shops and processing plants.

It has to be borne in mind that faulty meat handling, apart from affecting the quality and shelf-life of meat and processed products, may also endanger the health of consumers. Massive contamination will not only enhance meat deterioration caused by food-spoiling bacteria, but may also cause poisoning of consumers by toxin-producing micro-organisms. Toxin-producing micro-organisms find a favourable environment not only in fresh meat stored for a prolonged period at ambient temperatures, but also under certain circumstances in processed meat or even in preserved meat. Appropriate hygienic measures during slaughtering and meat handling are therefore indispensable.

A proper way of improving the slaughter hygiene is to carry out as much of the slaughter operations as possible with the carcass in a hanging position.

Rural Slaughtering on the Farm

Slaughter operations of small animals can easily be performed with the carcass in a hanging position. Problems may arise when slaughtering large animals such as cattle. The bleeding of the animal and cutting off the hind feet is done on the ground and the opening of the skin of the belly and leg region may also have to be done in this position. To complete skinning, evisceration and splitting, simple wooden structures and ropes to gradually hoist up the carcass are strongly recommended.

Remarkable improvements in slaughter hygiene can be achieved by this technique with much less visible contamination (dirt) and a lower degree of invisible contamination. The structure built for slaughtering can also be used for cutting and deboning the carcass.

Slaughterslabs

Slaughterslabs are used in rural communities where the throughput of animals is very small, but where slaughtering in the field should be avoided.

Animals should be suspended from a wooden or metal frame using a gantry hoist to ensure that carcasses are kept off the floor.

The simplest slaughterslab consists of a concrete platform with gantryhoist facilities. The platform should be approximately 50 cm above the ground and sloped to a drain. A roof should be added as protection against adverse weather conditions.

Slaughterslabs should be fenced and roofed and have facilities for hygienic slaughtering and adequate disposal of effluents. If pigs are to be killed a scalding vat should be provided.

The Animal Production and Health Division of the Food and Agriculture Organization of the United Nations (FAO) has developed a model project for village meat industry in which one of the main components is a smallscale modular slaughterhouse. The design incorporates the use of locally available construction materials and unsophisticated equipment. The slaughterhouse can be built in modules, adding units to the central slaughterhall for operations such as by-product utilization, meat preservation, processing and butchering. Designs have also been prepared for the construction of a meat market in order to facilitate the integration of production, processing and marketing.

It is intended that from the basic nucleus a number of small-scale industries be developed consecutively, e.g. meat preservation by low-cost technologies, traditional tanning, handicrafts based on hides, skin, bone and horns, etc. The objective is to increase employment, in particular for rural women, create market outlets for livestock products and increase the income of small producers.

The project could become a focus for farmer organization and a centre of technical assistance for producers. The organization of small producers is indispensable to ensure their participation and for the success and continuity of the project.

FAO is promoting the establishment of slaughterhouses of this type in some rural areas because this would increase the availability and quality of meat, improve the utilization of by-products, avoid rural migration and increase employment. The slaughterhouses may, therefore, constitute a tool for rural development, increasing the income level in the countryside and villages, thus improving the living conditions of the rural communities.

Rural slaughterhouses can also facilitate veterinary control of livestock (ante-mortem inspection) and carcass meat (post-mortem inspection). Slaughtering can be organized to follow a determined daily or weekly time schedule for which compulsory veterinary controls can be arranged. In addition, adequate technical facilities for efficient meat inspection can be provided, i.e. vertical position of the carcasses and special platforms for the personnel. Special hooks or inspection tables can be provided for the inspection of internal organs.

The detailed design for a small-scale modular slaughterhouse was developed by FAO. It includes designs, specifications, and schedule of quantities for a slaughterhouse and meat market suitable for small communities.

Provision is made for slaughter of all species: cattle (or buffalo), sheep, goats and pigs, though because of space limitations, concurrent slaughter of different species is not possible. The abattoir capacity will be dependent on the mix of animals being slaughtered. Daily throughputs of approximately 10 large stock (e.g. cattle) or 50 small stock (sheep, goats or pigs) or a combination thereof could be achieved with this design.

The facilities are divided into modules which can be combined as required to suit a particular location.

The following modules are included:

- Lairage
- Slaughter floor
- Chiller
- Tripe room
- Meat cutting and processing room
- Solid waste and blood disposal
- Hides and skin processing
- Effluent disposal
- Electric light and power
- Water supply
- Meat market

A possible overall abattoir layout based on these modules is shown in Fig. 3. This layout shows a typical arrangement for a facility designed to handle beef, small ruminants and pigs. Modifications for a large beef kill and/or the elimination of pig slaughter (e.g. for Muslim communities) are possible.

It is evident that when establishing slaughterhouses each country or even separate localities must adopt a solution incorporating special local conditions, locally available materials and manpower, etc.

The designs foresee procedures for slaughter of each species as follows:

Cattle

The animal is led into the bleeding area where it is restrained by a tether through the floor ring prior to stunning (using a captive bolt pistol). After stunning the animal is shackled by one leg and hoisted with a rope pulley block. The animal is then stuck and allowed to bleed in this position and the blood collected in a drum for disposal.

Once bleeding is complete the head can be removed and the animal lowered on to the cradle for dressing. It is also possible to dress the hanging animal. The feet are then removed, the skin opened up along the breastbone and the hide partially flayed. Leg hooks are attached and the carcass raised to a half-hoist position on the spreader. Flaying can then be completed and the hide removed. The paunch can then be removed to the inspection buggy and the red offal (including lungs if treated as edible) placed on hooks or the inspection table for inspection.

After inspection the carcass can be split and quartered, the quarters being individually hung on the low rail.

Once the carcass has been partially flayed and half-hoisted a second animal can enter the bleeding area.

Pigs

Pigs are first stunned in the stunning area then hoisted for sticking and bleeding and then transferred to the scald tank. After scalding for approximately five minutes at 60°C the carcass is removed to the scraping table. After scraping a gambrel can be inserted into the hind legs and the carcass transferred to the overhead rail for final scraping and evisceration. Once a pig is clear of the scraping table the next pig can be placed in the scalding tub.

Sheep and goats

These would be slaughtered and dressed on the rail in the pig area in a similar manner to pigs. The scraping table is removed to one side during processing of sheep and goats.

Carcass Handling

After slaughter carcasses should be chilled in chilling rooms. Chilled meat is a requirement for many methods of further processing. However, in rural areas of developing countries refrigeration facilities are generally lacking. Before cutting, carcasses should therefore be carefully examined for signs of taint caused by microbial spoilage.

Unchilled meat as a raw material is suitable for dried meat and certain meat products which undergo a heat treatment immediately after processing.

When dealing with hot carcasses, cuts and trimmings should be either consumed or processed (dried) on the day of slaughtering. It is obvious that under these conditions good hygiene during slaughtering and meat handling is of great importance for the quality of the final product. The higher the initial contamination, the faster the meat deterioration, especially under high ambiental temperatures.

In this context a new way of short-term meat preservation that could be especially beneficial for prolonged periods of handling of carcasses or meat cuts during transport, etc. should be mentioned. Meat surfaces are treated with organic acids such as acetic, lactic, citric, tartaric and ascorbic acid, as well as sodium sorbate. These compounds are from different natural foods and not toxic and may be used alone or in combination as dipping solutions or sprays on the surface of meat and meat products. Treatment of carcasses with these products has proved to be successful under conditions in developing countries. With an aqueous solution of 20 percent sodium sorbate, 5 percent sodium acetate and 5 percent sodium chloride sprayed on warm beef carcasses, the shelf-life of meat at 25° to 35°C was doubled. However, more work is needed on the subject, especially for tropical conditions.

Cutting

As whole carcasses of beef or pork are too large to be easily transported in one piece, they are split into sides or cut into fore- and hindquarters. The development of meat processing introduced the need for cutting quarters, halves or whole carcasses into smaller pieces which, according to their quality and market value, are used for culinary purposes and processing respectively.

The culinary meat in the form of primary cut is mostly sold to wholesale dealers who bone and cut it into sub-primals and finally into retail cut. The point at which wholesale cutting ends and retail cutting begins is not clearly defined. In the traditional, small, independent butchers' shops these operations may take place on a cutting table behind the counter in order to give the butcher the maximum opportunity to select the most suitable piece of meat and prepare it in a manner which is most likely to satisfy the consumer.

Meat for processing comprises parts of lower quality but also high-quality meat for the manufacture of special products like hams, smoked pork loins and dried meat. It can be divided into different classes according to the amount of fat and connective tissues.

There are several regional, country, and local differences in cutting the animal carcass into primary or retail cut and butchery practices vary depending on geographical location, tradition and habits, and demand for high-quality meat.

For the preparation of dried meat, two methods have been recommended. In both, carcasses are divided into two sides along the backbone. Each side is cut crosswise into two quarters which are divided into primal and retail cuts, and then boned and trimmed.

The pistola cutting system, with the side of beef quartered between the fifth and sixth ribs, allows the complete separation of all first-quality meat cuts in the hindquarters and loin regions. Fig. 4 illustrates primal cuts in the beef carcass.

After separation of the forequarters, the flank piece is removed by freeing the muscles of the abdomen from those of the proximal pelvic limb. Separation is completed by extending the cut down the side, parallel to and at a distance of some 20 cm from the backbone.

The hindquarter piece, which contains all the first-quality muscles, may be separated into the hind leg, rump and loin cuts by cutting and sawing directly across the *Bicepsfemoris* at a point just below the exposed point of the pelvic bone. The forequarter is similarly divided into two by removing the foreleg and blade cut from the thorax and neck.

Table 1 illustrates the distribution of primal cuts within the carcass.

Table 1. Pistola system Primal cuts expressed in percentage of cold carcass weight

Primal cut	*Percentage*
Flank and rib	14
Hind leg	29
Rump and loin	20
Foreleg and blade	15
Thorax and neck	22

According to the second method the carcass is split into sides, along the spinal column. After that, the sides are cut horizontally into two quarters: hindquarters and forequarters. The cut line runs immediately after the last rib, which excludes leaving any ribs on the hindquarters. The hindquarters are hooked by the Achilles tendon and the forequarters by the last two ribs.

In both cutting systems further division into commercial or retail cuts is usually necessary in order to separate the muscles or groups of muscles most suitable for drying from those which, by nature of their characteristics and retail value, should be sold fresh or processed by other methods.

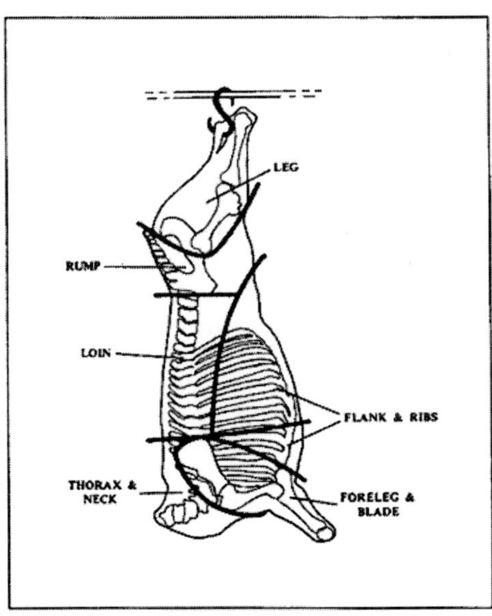

Figure 1. Beef carcass showing primal cuts according to pistola procedure.

After boning, the meat is trimmed to improve the quality of the finished product. Trimming consists of removing all or part of the extra fat layers, coarse tendons, fascia, nerves and blood vessels and, where necessary, bits of muscle which have been improperly cut or superficially damaged. Infiltrated fat, particularly melted fat, must also be removed to ensure a high-quality finished product which keeps well. Not only does this fat have a peculiar taste, it is sticky, unpleasant, easily becomes rancid and often forms a barrier which can interfere with the drying process.

The bones can be dried as well. They should be cut into smaller pieces on a clean clog. However, care should be taken to leave the connective tissue and ligaments intact, so that the pieces of the bones can be suspended for drying after having been dipped into a salt solution.

Some Important Non-meat Ingredients for Meat Processing

Salt is an important ingredient in the preparation of meat emulsions for imparting the typical flavour of processed meats and to contribute to keeping quality. The salt content of most processed meats ranges between 2.5 to 5.0 percent of the final product. A higher salt content would produce a salty taste. To function as a preservative, salt concentrations in the product of about 17 percent would be necessary, far too high for a palatable product.

Nitrates and nitrites are curing ingredients. The development of the red or pink colour is their most obvious effect. In addition, nitrites:

- affect the flavour of the products through their action as anti-oxidants; and
- have bacteriostatic properties, particularly in canned products (sodium nitrite is an effective inhibitor of the growth of the bacterium *Clostridium botulinum*).

Nitrate in itself is not effective in producing the curing reaction until it is first broken down into nitrite. This is a slow process which depends upon the presence of bacteria in brine and meat. Therefore nitrate has largely been replaced by nitrite.

Nitrite provides the ultimate source of the nitric oxide that reacts with the myoglobin pigment of the muscle tissue. Levels in excess of 200 parts per million of sodium nitrite should not be used. Since nitrate and nitrite are added in small amounts, it is recommended that they are first dissolved in water to ensure uniform distribution.

Extenders in general are added to lower-quality products for economic reasons. Some of them improve binding properties, cooking yields, slicing characteristics and flavour. The ones most often used are:

- dried skim milk and milk proteins;
- various cereal flours such as wheat, rice, oats, corn;
- soy products such as flours (containing about 50 percent protein), grits (which are similar to soy flour in composition but they are larger in particle size and more adaptable to meat products), textured soy protein (similar to grits except that the texture is changed to more closely resemble the texture of ground meat), soy protein concentrates (containing about 70 percent of protein, that is available either in a coarse granular form or as a flour), soy protein isolates (containing about 90 percent protein), useful both as binders and emulsifiers.

Various non-meat proteins are being developed for use in sausage and processed meats. These include protein from oil seeds other than soya, such as cotton seed and peanuts, as well as from single-cell plant source such as torula yeast. Most extenders are usually limited to 3 percent in the dry state or to 10 percent after swelling.

Seasoning is a comprehensive term for ingredients which improve the flavour of processed meats. Salt and pepper form the foundation upon which many seasoning formulae are built. Other ingredients such as spices, herbs and vegetables are supplementary, although necessary to obtain the distinctive flavour associated with various products.

Spices are aromatic substances of vegetable origin and include cinnamon, cassia, clove, ginger, mace, nutmeg, paprika, pepper, cardamom, coriander and mustard. They vary in composition. The aromatic and pungent components which render them valuable are present in volatile oils and resins. Some success has been achieved in extracting these components. Spice extracts offer the advantage of being easier to dose and to store.

Condiment herbs include sage, savory, bay leaves, thyme and marjoram. The dried leaves of any of them can be used in the preparation of sausages and other meat products. Condiment vegetables are onion and garlic.

All spices and seasonings should be stored under dry conditions and possibly in sealed containers. They should not be exposed to direct sunlight and should only be ground or crushed on the day of manufacture.

Simple Techniques for Production of Dried Meat

Principles of Meat Drying

Drying meat under natural temperatures, humidity and circulation of the air, including direct influence of sun rays, is the oldest method of meat preservation. It consists of a gradual dehydration of pieces of meat cut to a specific uniform shape that permits the equal and simultaneous drying of whole batches of meat.

Warm, dry air of a low humidity of about 30 percent and relatively small temperature differences between day and night are optimal conditions for meat drying. However, meat drying can also be carried out with good results under less favourable circumstances when basic hygienic and technological rules are observed. Intensity and duration of the drying process depend on air temperature, humidity and air circulation. Drying will be faster under high temperatures, low humidity and intensive air circulation.

Reducing the moisture content of the meat is achieved by evaporation of water from the peripheral zone of the meat to the surrounding air and the continuous migration of water from the deeper meat layers to the peripheral zone.

There is a relatively high evaporation of water out of the meat during the first day of drying, after which it decreases continuously. After drying the meat for three or four days, weight losses of up to 60–70 percent can be observed, equivalent to the amount of water evaporated. Consequently, moisture losses can be monitored by controlling the weight of a batch during drying.

Continuous evaporation and weight losses during drying cause changes in the shape of the meat through shrinkage of the muscle and connective tissue. The meat pieces become smaller, thinner and to some degree wrinkled. The consistency also changes from soft to firm to hard.

In addition to these physical changes, there are also certain specific biochemical reactions with a strong impact on the organoleptic characteristics of the product. Meat used for drying in developing countries is usually derived from unchilled carcasses, and rapid ripening processes occur during the first stage of drying as the meat temperature continues to remain relatively high. For that reason the specific flavour of dried meat is completely different from the characteristic flavour of fresh meat. Slight oxidation of the meat fats contributes to the typical flavour of dried meat.

Undersirable alterations may occur in dried meat when there is a high percentage of fatty tissue in the raw meat. The rather high temperatures during meat drying and storage cause intensive oxidation (rancidity) of the fat and an unpleasant rancid flavour which strongly influences the palatability of the product.

Meat drying is a complex process with many important steps, starting from the slaughtering of the animal, carcass trimming, selection of the raw material, proper cutting and pre-treatment of the pieces to be dried and proper arrangement of drying facilities. In addition, the influence of unfavourable weather conditions must also be considered to avoid quality problems or production losses. The secret of correct meat drying lies in maintaining a balance between water evaporation on the meat surface and migration of water from the deeper layers.

In other words, care must be taken that meat surfaces do not become too dry while there is still a high moisture content inside the meat pieces. Dry surfaces inhibit the further evaporation of moisture, which may result in products not uniformly dried and in microbiological spoilage starting from the areas where the moisture content remains too high.

Adherence to the following meat-drying techniques should avoid failures in the production of dried meat and ensure obtaining products of good quality with a long shelf-life. The following description of the basic technology of meat drying includes the salting of the meat before drying. Presalting is not absolutely necessary, but has certain advantages, particularly for the drying of meat strips and large flat meat pieces and is therefore strongly recommended for this type of product.

Selection of Meat for Drying

As a general rule only lean meat is suitable for drying. Visible fatty tissues adhering to muscle tissue have a detrimental effect on the quality of the final product. Under processing and storage conditions for dry meat, rancidity quickly develops, resulting in flavour deterioration.

Dry meat is generally manufactured from bovine meat although meat from cameloids, sheep, goats and venison (e.g. antilopes, deer) is also used. The meat best suited for drying is the meat of a medium-aged animal, in good condition, but not fat. Meat from animals in less good nutritional condition can also be used for drying, but the higher amount of connective tissue is likely to increase toughness.

It is very important that raw material for the manufacture of dry meat is examined carefully for undersirable alterations such as discoloration, haemorrhagic spots, off-odours, manifestation of parasites, etc. Such defects must be trimmed off.

Carcasses have to be properly cut to obtain meat suitable for drying. Owing to their size, beef carcasses are more difficult to handle under rural conditions than carcasses of sheep, goats or game. In the absence of chilling facilities, beef carcasses must be cut and deboned immediately after slaughter.

Beef Carcass Cutting, Trimming and Deboning

Carcass cutting

The carcass is first split into two sides along the spinal column and then cut into quarters. Fore- and hindquarters are separated after the last rib, thus leaving no ribs in the hindquarter. For suspension the hindquarter is hooked by the Achilles tendon and the forequarter by the last two ribs.

Trimming

After the quarters are suspended so that they do not touch the floor or anything around them, they are trimmed. Careful trimming is very important for the quality and shelf-life of the final product. The first step is to remove with a knife all visible contamination and dirty spots. Washing these areas will spread bacterial contamination to other parts of the meat surface without cleaning the meat.

After completing the necessary cleaning of the meat surfaces, knives and hands of personnel must be washed thoroughly. Using a sharpened knife, the covering fat from the external and internal sides of the carcass and the visible connective tissue, such as the big tendons and superficial fasciae, are carefully trimmed off.

Deboning

It is recommended that this operation should start with the hindquarters and follow with the forequarters. The aim is to remove the bones with the least possible damage to the muscles. Incisions into the muscles are inevitable but only at spots where the bones adhere and have to be cut off.

Deboning of the suspended hindquarter should start from the leg and proceed to the rump and muscles along the vertebral column.

Deboning of the forequarter must start with cutting and deboning the shoulder separately, followed by cutting off the rib set, together with the intercostal muscles.

Deboning of the forequarter is completed by removing the meat from the neck and the breast region of the spinal column.

Technique of Cutting Meat Pieces for Drying

Anatomic cuts, which were separated from the carcass, are suspended again and the big individual muscles are carefully cut out, while the smaller muscles are left together. The next step consists in cutting the muscles into thin strips. This operation is crucial for the appearance and quality of the final product. All strips to be dried in one batch must be cut to an identical shape. Care must also be taken to obtain rather long strips of meat.

There are two ways of cutting muscles or smaller muscle groups into strips:

- cutting the meat after placing it on an appropriate clean chopping board; or
- cutting the muscle in the hanging position.

In both cases the muscles have to be split exactly along the muscle fibres. The strips must be cut as uniformly and as smoothly as possible and the diameter of the strip must remain the same throughout the length.

The length of the strips may differ, though it should not be less than 20 cm and not more than 70 cm. Meat cut into shorter strips requires considerably more time for hooking than the same quantity cut into longer strips. However, strips which are too long may break because of their weight.

Beef muscles suitable for drying are usually no longer than 50 cm (except the sirloin strip attached to the spinal column). However, strips longer than 50 cm can be produced by cutting the muscle along the fibre in one direction, without cutting through the end of the muscle. Using this technique long strips can be obtained, but their length should not exceed 70 cm for reasons of stability. The thickness of the strips determines the duration of the drying process. Since thick strips take considerably more time to dry than thin ones, it is important that strips to be placed in the same batch are of the same cross-section, with only the length differing. Insufficiently dried or overdried pieces will be the result if this rule is not followed.

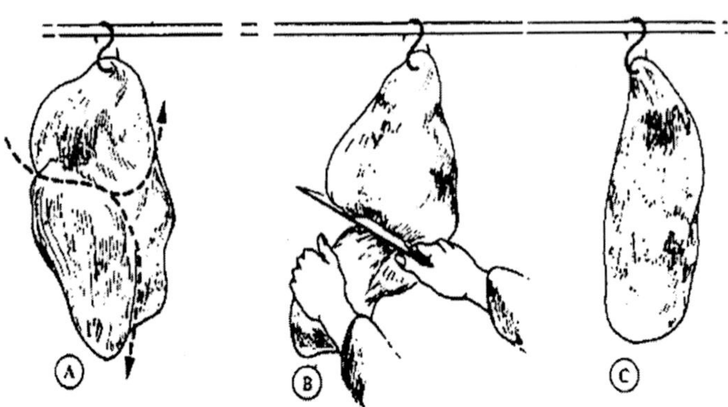

Figure 2. Suspended anatomic cut from the hindquarters (A) and splitting into individual muscles (B) which result in (C).

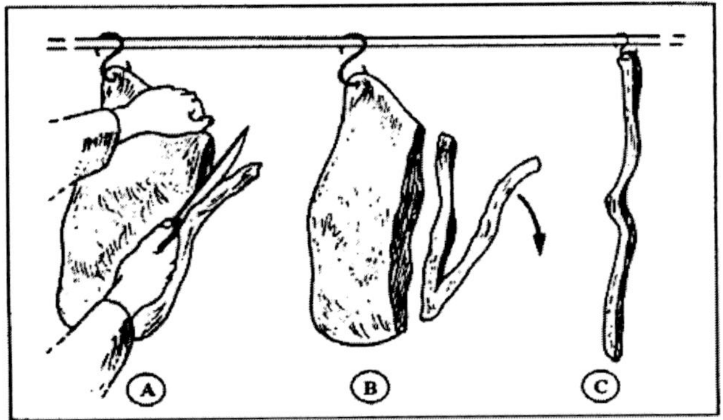

Figure 3. Special cutting technique to obtain long meat streeps

Cutting muscles into long, thin and uniformly shaped strips requires experience and skill. Knives with broad blades are best suited for this purpose.

Under dry climatic conditions two basic shapes of meat pieces proved to be the most suitable for natural drying:

- strips with a rectangular cross-section of 1 x 1 cm; and
- flat-or leaf-shaped pieces with cross-sections of max. 0.5 cm x approx. 3, 4 or 5 cm.

Recommended Treatment before Drying

Because meat is always consumed slightly salted, the raw material may be presalted before drying. This procedure not only contributes to a more tasty product, but is also desirable from the technological and hygienic standpoint. Pure common salt is used for this purpose, either dry or dissolved in water. In the case of meat for drying cut into strips or flat pieces, the use of a 14-percent salt solution is preferred.

Dipping the meat into the salt solution serves first of all to inhibit microbiological growth on the meat surfaces. For that reason salting has to be carried out within five hours after slaughter, as after that period massive microbiological growth occurs which cannot be reduced by salt treatment. Secondly, presalting is a protection against insects during drying. The freshly cut meat surfaces are very attractive to various insects, in particular domestic flies, which feed on the moisture excreted from muscle fibres. These insects cause considerable contamination of the meat and may also deposit their eggs into it. Meat is no longer such an attractive environment for insects after it has been dipped into the salt solution. The salt concentration on the meat surfaces keeps them away.

Furthermore, a thin layer of crystalline salt is formed on the surface of the meat during drying. The salt crystals are hygroscopic and absorb part of the water excreted from the meat, preserving the meat surfaces by keeping them dry. Dry meat surfaces inhibit the growth of bacteria and moulds which is one reason for the preservability of presalted and dried meat.

The salt solution is prepared by adding the necessary amount of edible common salt to water and dissolving it by intensive stirring. To obtain the recommended salt concentration of about 14 percent the amount of salt necessary for different volumes of water (expressed in litres) is indicated below:

Water (1)	*Salt (g)*
5	810
6	975
7	1 140
8	1 300
9	1 460
10	1 630

As soon as the salt is dissolved in the water, the meat strips are dipped into the solution, soaked for about five minutes and then drained. Draining should be done by placing the strips into a plastic sieve in order to allow the brine to drop off for collection and re-use.

The handling of the meat strips before drying has to be carried out under strictly clean conditions in order to avoid contamination and ensure a long shelf-life of the dried product. However, if accidental contamination of certain pieces occurs, further processing can only be undertaken with certain precautions. A special bucket with salt solution should be available in order to soak the contaminated pieces of meat, after having rinsed them previously in clean water. However, it must be borne in mind that the original quality of contaminated pieces cannot be restored. For that reason such pieces should always be dried separately, and not stored for a long period, but should be used as soon as possible in the preparation of meals.

Methods of Suspending Meat Strips for Drying

The traditional way of suspending meat for drying by hanging strips over tree branches, wire or rope is not recommended because meat remains in contact with these supporting devices or may touch each other and thus not dry properly in these contact areas. Consequently, the chosen method should be to suspend the meat strips individually from one end, thus ensuring, through appropriate arrangement on the drying facility, free air circulation along the whole length of the pieces and fast and uniform drying. The contact of meat pieces with each other during drying must absolutely be avoided, since these areas will remain wet and humid for a prolonged period, thus making them a favourable environment for spoilage, bacteria and flies.

Suspension using metal hooks

This is a very simple but efficient way of suspending the meat strips. The meat strips are hooked at one end, always the thicker end for stability, and suspended on a horizontal wooden stick, tightrope or wire.

The metal hooks can easily be made, preferably from galvanized (non-corrosive) wire. Wire of 1 to 1.5 mm diameter is cut into pieces 15 cm long with a slanting cut so that the ends are sharp to allow piercing of the meat. In order to obtain an S-shaped hook, both ends of each piece are simply twisted around a circular stick.

For this purpose a thin string or a somewhat stronger thread is best suited. The string is divided into pieces about 30 cm long with the ends knotted. The string is fixed to the stronger end of the meat strip by a double loop and pulled tight in order to prevent the meat from slipping out of the loop.

Suspension using metal clips

Clips 4 to 7 cm wide are best suited. They are easily placed on the stronger end of the meat strips. Whereas metal hooks and rope loops can only be used for the suspension of rectangular or similar shaped strips, the metal clips are very practical for the suspension of flat, leaf-shaped pieces. The special advantage of hanging leaf-shaped pieces by means of a clip is that the edges of the meat do not fold in during drying.

Installation for Drying Entire Batches of Meat

It has already been pointed out that placing meat pieces for drying over wire, ropes or branches of trees is not recommended. Apart from problems of free air circulation under trees, some pieces may be intensively exposed to direct sun, whereas others are screened by the foliage. Furthermore, wind will transfer dust, twigs or leaves on to the meat and insects and birds will cause further damage. A general disadvantage of this very simple method of meat drying is that it is practically impossible to shelter the meat in case of storm or rain.

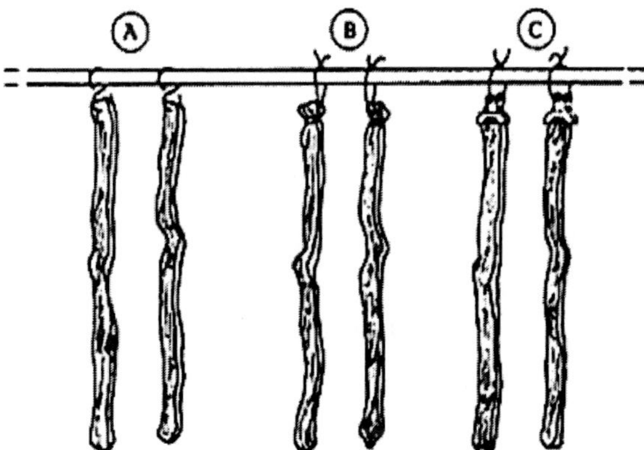

Figure 4. Suspension of meat strips on hooks (A), loops (B), and by means of clips (C).

It is therefore recommended that natural meat drying be done by using simple premises and equipment, which can be made locally. The following descriptions give the different types of meat dryers.

These meat dryers are constructions of wood, metal and/or concrete, stationary or mobile, without or with a roof. For strips suspended by hooks or with a loop attached or fixed by clips, removable horizontal bars, either made of wood or metal or horizontal wire strings are needed.

Sun meat dryer made of wood or metal

This dryer consists of four wooden forks planted into the ground which are connected with two longitudinal, wooden traverses of about 4 m. Wooden or metal sticks for hanging the meat streeps are placed on the traverses at a distance of 15 cm from each other. If there are not sticks available, strong metal wire or plastic rope can be spanned between the two longitudinal traverses and two additional transversal traverses should be fitted for reinforcement. Similar constructions, but with iron parts and traverses instead of the wooden ones are suited for the more industrial type of meat drying. These constructions can also easily support a roof and meat drying can be done on two or more levels. However, care has to be taken that the traverses at the lowest level are not less than 1 m from the ground.

Mobile meat dryer

This type of dryer, which can be easily assembled or dismantled, can be moved to places where the animals are slaughtered. Although wooden constructions can be used for this purpose, for easy assembly and for a firm base metal constructions using 40-mm tubular iron bars are more convenient.

This type of dryer consists of two rectangular frames (2 x 2 m) placed 4 m apart from each other. They are connected with four longitudinal and eight transversal metal traverses in two levels from the ground (approx. 1 m and 2 m), thus permitting meat drying on the two levels.

This dryer has the capacity for drying the meat of two beef carcasses at the same time. It is recommended that the upper level of the dryer be used for the suspension of meat from the hindquarters and the lower level for meat from the forequarters.

Meat dryer with protection against external influences

In regions with strong and frequent winds, meat placed on the dryer must be protected from contamination by dirt, dust, sand, etc. In these cases it is recommended that the side walls of a roofed dryer be covered with plastic foil up to a height of 0.80 m to 1.20 m from the floor. It is important to ensure that the upper parts of the dryer remain open for air circulation.

Protection against insects is provided by covering the sides of the dryer with insect screen.

Arrangement of the Meat Strips in the Dryer

In the dryer the meat strips are hung on horizontal plastic ropes, wires or sticks by means of hooks, loops or clips. As shown above, ropes and wires have often to be supported because of the heavy load of suspended meat, and they are not easily removable, which may cause difficulty in handling the meat to be dried.

Compared to flexible wires or ropes, firm wooden or metal sticks have proved to be the better solution for the following reasons:

- the sticks can be moved out of the dryer in order to suspend the meat pieces;
- the suspended meat pieces maintain the correct distance from each other on the firm stick and there is no need for supports for the sticks;
- sticks with the suspended meat can easily be moved, for example to a smoke house or for shelter in case of a storm or heavy rain; and
- weight losses can easily be monitored during drying by weighing the stick plus the meat without being forced to remove the meat strips.

The round wooden sticks can be made locally; they are usually 2 m long and 2 cm in diameter. This size permits the suspension of 25 to 30 pieces of meat. Metal sticks are in two different shapes, one tubular, made of galvanized water-pipes, and the other T-shaped. They are more expensive than wooden sticks but they last much longer.

The following rules for the arrangement of the meat pieces apply to both wooden or metal sticks:

- the number of meat pieces suspended must always remain the same for all sticks in the dryer (for instance 30 per stick), for reasons of optimal air circulation and easy control of theft, etc.;
- the distance between the individual suspended pieces must remain the same and be sufficient for free air circulation; and
- the longest and thickest pieces of meat have to be placed toward the ends of the stick whereas the thinnest and shortest pieces are kept in the middle, in order to expose the larger pieces to the stronger air circulation on the external part of the dryer.

Quality of the Finished Product

The appearance of the dried meat should be as uniform as possible. The absence of large wrinkles and notches indicates the desired steady and uniform dehydration of meat.

The colour of the surface, as well as of the cross-cut, should be uniform and dark red. A darker peripheral layer and bright red colour in the centre indicates incorrect, too fast drying, with the formation of hard rind which hinders evaporation from the deeper layers of the product. In this case the central parts have a brighter colour and softer consistency and are, because of the higher water content, more susceptible to microbiological spoilage when packaged or otherwise stored. A softer consistency can also be recognized by pressing the meat with the fingers. These pieces should be kept for one more day in the dryer for finishing. The consistency of properly dried meat must be hard, similar to frozen meat.

Taste and flavour are very important criteria for the acceptance of dried meat by the consumer. Dried meat should possess a mild salty taste which is characteristic for naturally dried meat with no added spices. Off-odours must not occur. However, a slightly rancid flavour which occurs because of chemical changes during drying and storage is commonly found in dried meat. Dried meat with a high fat content should not be stored for a long period but used as soon as possible in order to avoid intensive rancidity.

Dried meat must be continuously examined for spoilage-related off-odour, which is the result of incorrect preparation and/or drying of the meat. Meat with signs of deterioration must be rigorously sorted out.

Packaging and Storage

After taking the dried meat strips out of the dryer, a selection of the pieces based on length can be undertaken.

Packaging serves to protect the product from contamination to which the meat might be exposed on its way from the producer to the consumer.

Numerous materials are used for packaging dry meat, such as paper, plastic foils, aluminium foils, cellophane and textiles. The longest shelf-life is obtained using vacuum-packaging. Transparent plastic material and cellophane are more appealing to the consumer.

Packaging is employed for both the retail and wholesale trade. The weight per package of dry meat for retail sale usually does not exceed 1 kg, whereas those for the wholesale trade weigh 5, 10, 25, or 50 kg.

If plastic bags are used for packaging, the pieces of dry meat should be cut to a certain length so that they can be best arranged in the bags. Cardboard boxes are very useful for additional packaging.

During storage special care has to be taken to prevent dried meat, which is not packaged in water-proof containers, from becoming wet, resulting in rapid growth of bacteria and moulds. For this reason the premises for storing dry meat have to be rain-proof. It is further advisable to cover the piles of packaged dry meat with plastic sheets, as additional protection against moisture and dust. Dry meat protected in this way can be stored for more than six months.

During storage individual packages must be opened at least once a month and the organoleptic quality of the goods examined. These controls enable the persons responsible to evaluate storage conditions and to assess the shelf-life of the dry meat.

For controlling temperature and air humidity, it is useful to have a thermometer and hygrometer installed on the premises. A maximum-minimum thermometer is recommended to obtain the highest and lowest temperatures recorded between two readings. The temperature and relative air humidity should be carefully registered bearing in mind that dry meat is extremely sensitive to changes in environmental conditions, especially of the ambient temperature and relative humidity.

Preparation of Dried Meat for Consumption

Dried meat manufactured as described above has to be rehydrated to resemble fresh meat again. Rehydrated dried meat has almost the same nutritive value as fresh meat. Rehydration is in most cases combined with cooking. The procedure usually starts by putting the dried meat, which may be cut in smaller pieces, into a pot. The meat in the pot is then covered with water and boiled. The rehydrated and cooked meat and the broth are used, together with other additives which may vary according to local consumption habits, for the preparation of tasty dishes. Other types of dried meat, which are manufactured by a combination of drying with special treatments, are consumed raw, without rehydration and cooking.

Meat Drying in Combination with Additional Treatment

Meat drying after presalting, as described above, is the simplest and most efficient method of meat dehydration. Additional treatments used for some special dried meat products are curing, smoking and the utilization of spices and food additives.

Specific antimicrobial agents in smoke or spices or the antimicrobial properties of the curing substance, nitrite, may allow a less intensive dehydration of the meat. The resulting "semi-dry" products are in most cases consumed without rehydration, whereas rehydration is indispensable for common dried meat. In many countries, including developed countries, "semi-dry" products such as unsmoked and smoked raw hams (e.g. Parma ham, jamon serrano or smoked hams of the central European type), unsmoked or smoked dry sausages (e.g. salami, dry chorizo) or dried cured beef (Bündnerfleisch of Switzerland) are not only popular because of the products' durability but particularly because they are delicious, high-quality meat specialities.

In developing countries, where the preservation aspect is even more important because of the lack of a cold chain, treatment carried out in addition to the drying of meat will be somewhat different and in some cases (e.g. intensive smoking over fire) the product quality is lowered rather than improved. The reasons for this additional treatment are in many cases adverse climatic or environmental conditions which do not allow the drying of meat without additional treatment. There are also of course other reasons for additional treatment, such as special flavours or special mixtures with non-meat ingredients, which may be preferred locally.

Cured dried meat

Curing is the impact of nitrite on meat, in particular on the muscle pigment, myoglobin, which results in the formation of the pigment myochromogen and gives a stable red colour to muscle tissue. In addition, nitrite inhibits to some extent microbiological growth in the meat, but does so efficiently only in combination with low temperatures and/or low water activity. These effects are of particular importance for the shelf-life of raw hams and dry sausages and may also be of importance for non-intensively dried biltong, the South African dried meat, which may also be manufactured with nitrite or nitrate.

Apart from occasional use in biltong, it can be concluded that curing is not important in the manufacture of traditional dried meat products. The reasons are that a bright red colour is not desired in dried meat (because it will be rehydrated and used for cooking meals) and drying is generally so intensive that the inhibiting effect on microbiological growth is unnecessary. Curing substances must be handled very carefully as they are toxic even in low concentrations. Very small dosages are sufficient for the curing effect, about 200 ppm, that is, 2 g or less in 10 kg meat.

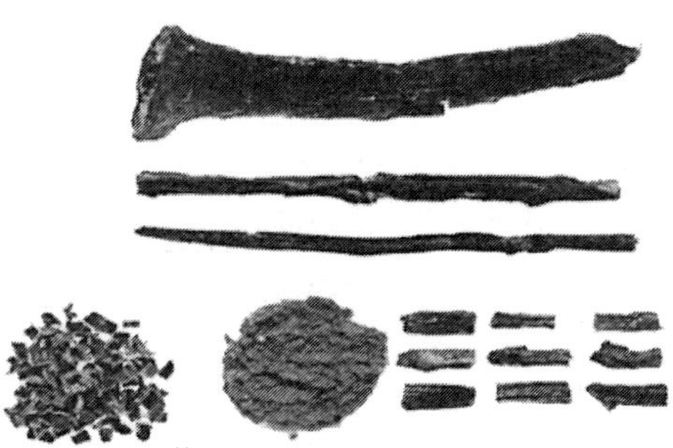

Figure 6. Whole strips and flat pieces of dried meat and dried meat comminuted to fragments of different sizes for preparing meals.

Smoked dried meat

Smoking of meat is a technique in which meat is exposed directly to wood smoke which may be generated by a variety of methods. In smoke produced from wood there are various substances which contribute to the flavour and the appearance of the smoked meat product and which have a certain preserving effect on the product.

However, the preserving effect of common smoking is not very significant when storing the product without a cold chain. On the other hand, intensive or prolonged smoking may considerably increase the shelf-life of the product, but it also has an unfavourable effect on flavour. Whereas a light smoke aroma generally enhances the organoleptic properties of the product, intensive smoking has a negative influence on the quality, especially in the case of prolonged storage in which concentrated smoke compounds develop increasingly unpleasant tarry flavours.

In view of the above, smoking in order to preserve meat can only be considered as an emergency measure when no other preservation methods can be carried out. This may be the case during wet weather or generally under a humid climate, or when the preservation has to be completed as fast as possible because of the need of immediate transport, for instance after game-hunting.

Intensive meat smoking is always a combination of two effects, drying the meat by reducing its moisture content through hot air and the condensation of smoke particles on the meat surface together with their penetration into the inner layers of the product. Both have preservative effects and prolong the shelf-life of the product.

To smoke the meat, large strips and/or pieces, with and without bones, are dried by smoking in special drying/smoking places. The smoke is produced in these cases by glowing wood. Often, meat is prepared quickly by drying and smoking over a fire. In this case, the meat is not only smoked, but "half-cooked" or roasted. Normally, meat from this treatment is not well prepared and has to be consumed soon after drying, otherwise it will spoil quickly.

The quality of traditionally smoke-dried meat is generally poor. This is not only owing to poor meat quality or inadequate smoking devices, but mainly because smoke-drying is a rather rough treatment for the meat. The process is fast and has a certain preserving effect, but at the cost of quality.

Quality losses are even more obvious when failures in preparing the raw material occur. When, for example, the thickness of the meat parts to be smoked ranges from about 3 cm to 15 cm, uniform drying will not be achieved. The smaller pieces will be overdried and the thicker ones may still remain with a high moisture content in the product centre. The results of faulty drying and smoking are a too strong smoke flavour, lack of rehydration capacity of the smaller parts and fast spoilage of the thicker parts. For effective smoke-drying, the meat thickness should not exceed 7 cm to achieve products which are stable for a certain period without refrigeration.

Apart from primitive smoking places with just a fire below the meat, the construction of special smoking kilns has been suggested for smoke-drying of meat.

The effect of light smoking could be of interest for the production of dried meat. Light smoking is not suitable for meat preservation without a cold chain, but it adds a smoke flavour to the product and inhibits the growth of moulds and yeasts on the product's surface owing to the fungistatic smoke compounds. Thus light smoking may be used for the prevention of growth of moulds during the storage period of dried meat, especially under humid climatic conditions.

Dried meat with spices and additives

Various methods, typical for different regions, exist to produce this type of product. General guidelines for manufacture cannot be given because of the great variety of preparations, but the idea behind all of them is to combine the necessary preservation of meat with a typical flavour. In some cases the additives act as an absorbent with the aim of faster drying, and some spices may also act against bacterial growth.

Typical Dried-meat Products With or Without Additional Treatment

Some examples of dried meat or dried and further processed meat manufactured in Africa, America and the Near East are described. No mention is made of the various dehydrated meat products well known in the Far East. These products are somewhat different owing to their sugar component.

Odka (Somalia and other East African countries)

Odka is basically a sun-dried meat product made of lean beef and is of major importance to nomads in Somalia. In the face of perennial incidence of drought in the Horn of Africa, odka has become important since it is often prepared from drought-stricken livestock.

The production of odka is similar to the simple drying technique described earlier. However, the meat strips cut for drying are bigger and dry salting is usually applied instead of brine salting. After only four to six hours' sun-drying the large pieces of meat are cut into smaller strips and cooked in oil. After this heat treatment drying is continued and finally sauces and spices are added. For storage odka is again covered with oil and, when kept in a tightly closed container, it has a shelf-life of more than 12 months.

Qwanta (Ethiopia and other East African countries)

Qwanta is manufactured from lean muscles of beef which are further sliced into long strips ranging from 20 to 40 cm and are hung over wire in the kitchen to dry for 24 to 36 hours. Prior to drying, the strips are coated with a sauce containing a mixture of salt (25 percent), hot pepper/chilli (50 percent) and aromatic seasoning substances (25 percent). After air drying the meat pieces may be further exposed to a light wood smoke and are then fried in butter fat and dried again to some extent. At this stage the product is ready for consumption or storage.

Kilishi (Nigeria and other arid or semi-arid zones of West Africa)

Kilishi is a product obtained from sliced lean muscles of beef, goat meat or lamb and is made on a large scale under the hot and dry weather conditions prevailing from February to May. It is produced by sun-drying thin slices of meat. However, recent experience indicates that kilishi can also be produced industrially using tray-drying in a warm air oven. Connective tissue and adhering fatty material are trimmed off the meat which is cut with a curved knife into thin slices of about 0.5 cm thickness, 15 cm length and as much as 6 cm width.

Traditionally, the slices of meat are spread on papyrus mats on elevated platforms or tables in the sun for drying. However, these papyrus mats may lead to hygienic problems, especially after repeated use. Therefore, easily washable corrosion-free wire nets or plastic nets are recommended for horizontal drying. The vertical drying method is also recommended in this case.

Sun-drying of kilishi could also be improved by the use of solar dryers. These devices will increase the rate of drying of the product and keep insects and dust from the product.

In the first stage of drying, which takes two to six hours, the moisture of the meat slices has to be reduced to about 40 to 50 percent. The slices are then put into an infusion containing defatted wet groundnut cake paste or soybean flour as the main component (about 50 percent), and is further composed of water (30 percent), garlic (10 percent), bouillon cubes (5 percent), salt (2 percent) and spices such as pepper, ginger and onion. The "dried" slices of meat should absorb the infusion up to almost three times their weight.

After infusion, the wet product is again exposed to the sun to dry. Drying at this stage is much faster than at the first stage. When the moisture content of the slices has been reduced to 20 to 30 percent, a process which takes two to three hours depending on weather conditions and the dimensions of the product, the slices are finally roasted over a glowing fire for about five minutes. The roasting process helps to enhance desirable flavour development and to inactivate contaminating micro-organisms. Roasted kilishi is therefore superior in flavour to the unroasted version.

After roasting, the final moisture content ranges between 10 to 12 percent. It will decrease during storage at room temperature to as low a level as 7 percent. When packaged in hermetically sealed, low density plastic bags the product remains remarkably stable at room temperature for a period of about one year.

Biltong (Southern African countries)

Biltong is a well-known salted, dried meat prepared from beef or antilope meat. Most muscles in the carcass may be used but the largest are the most suitable. The finest biltong with the best flavour is made from the sirloin strip and the most tender is derived from the fillet.

The meat is cut into long strips (1 to 2 cm thick) and placed in brine, or dry-salted, which is actually the most popular method. Common salt, preferably coarse salt (1 to 2 kg for 50 kg of meat), or salt and pepper are the principal ingredients used, although other ingredients such as sugar, coriander, aniseed, garlic or other spices are included in some mixtures to improve flavour. In most cases nitrate or nitrite is added to achieve a red colour and the typical flavour of cured meat. The addition of 0.1 percent potassium sorbate to the raw meat is permitted in South Africa as a preservative. The salt/spice mixture is rubbed into the meat by hand and the salted strips are then transferred to a suitable container. It is recommended that a little vinegar be sprinkled on each firmly packed layer in the container.

Biltong is left in the curing brine for several hours, but not longer than 12 hours (otherwise it will be too salty), and then dipped into a mixture of hot water and vinegar. The biltong is now ready for sun-drying for one day. Then the strips are moved into the shade for the rest of the drying period. The product is usually not smoked, but if it is smoked only light cold smoking is recommended, which takes one to two weeks under sufficient air circulation. The biltong is ready when the inside is soft, moist and red in colour, with a hard brown outer layer.

Biltong is sold in sticks or slices. The usual shelf-life is several months without refrigeration and packaging, but in airtight packages the product stores well for more than one year. Biltong is not heated during processing or before consumption. It is eaten raw and considered a delicacy.

Pastirma (Turkey, Egypt, Armenia)

Pastirma is salted and dried beef from not too young animals. In some areas camel meat is also used. The meat is taken from the hindquarters and is cut into 50 to 60 cm long strips with a diameter of not more than 5 cm. The strips are rubbed and covered with salt and nitrate. The dosage of the nitrate in relation to the meat is 0.02 percent, that means 2 g of nitrate for 10 kg of meat. Several incisions are made in the meat to facilitate salt penetration.

The salted meat strips are arranged in piles about 1 m high and kept for one day at room temperature. They are turned over, salted again, and stored in piles for another day. Thereafter the meat strips are washed and air-dried for two to three days in summer and for 15 to 20 days in winter. After drying the strips are piled up again to a height of 30 cm and pressed with heavy weights for 12 hours. After another drying period of two to three days the meat pieces are again pressed for 12 hours. Finally the meat is again air-dried for 5 to 10 days.

After the salting and drying process, the entire surface of the meat is covered with a layer (3 to 5 mm thick) of a paste called cemen, which consists of 35 percent freshly ground garlic, 20 percent helba (i.e. ground trefoil seed), 6 percent hot red paprika, 2 percent mustard, and 37 percent water. Helba is used as a binder of the paste; the other ingredients are spices, but garlic is the most important as it is antimycotic. The meat strips covered with cemen are stored in piles for one day, and thereafter are dried for 5 to 12 days in a room with good air ventilation, after which the pastirma is ready for sale. Thus, the production of pastirma requires several weeks. However, not much energy is required since most of the salting and drying is done at room temperature. The final product has an average water activity (a_w) of 0.88. The a_w should not fall below 0.85 or the meat will be too dry. The salt content should range between 4.5 and 6.0 percent. The product is mould-free for months at ambient temperature even in summer. Pastirma thus has a better microbiological stability than biltong.

Charque (Brazil and other South American countries)

Charque consists of flat pieces of beef preserved by salting and drying. The fresh, raw meat from the fore- and hindquarters is cut into large pieces of about 5 kg, which should not be more than 5 cm thick. The pieces are submerged in a saturated salt solution for about one hour in barrels or cement vats. On removal from the brine, the meat is laid on slats or racks above the brine tank to drain.

For dry-salting, the flat meat pieces are piled on a sloping, grooved, concrete floor under a roof. To form a pile, salt is spread evenly over the floor about 1 cm high. Then a layer of meat is put on the salt. The meat is covered with another (1 cm) layer of salt followed by adding another layer of meat, and so on until the alternate layers of salt and meat reach a height of about 1 m. The pile is then covered with a few wooden planks and pressed with heavy stones.

After eight hours the pile is restacked so that the top meat goes to the bottom of the pile. The restacking process with fresh layers of salt is repeated every day for five days.

The salted meat is then ready for drying. Before initiating drying, the meat pieces are subjected to rapid washing to remove excess salt adhering to the surface. The meat pieces may also be passed through a pair of wooden rollers or a special press to squeeze out some surplus moisture and flatten the meat slabs. The meat is then spread out on bamboo slats or loosely woven fibre mats in a shed or, in industrial production, exposed to the sun on wooden rails which are oriented north-south, thus permitting an even solar coverage.

Initial drying, directly in the sun, is limited to a maximum period of four to six hours. This period of exposure may be subsequently lengthened to a maximum eight hours. Temperatures in excess of 40°C on the meat surface should be avoided. To ensure even drying over the extended muscle pieces, the meat is placed on the rails during the morning and removed again in the afternoon. The meat pieces are exposed to the sun each day over a period of four to five days. After each period of exposure the pieces are collected, stacked in piles on concrete slabs and covered with an impermeable cloth to protect them against rain and wind and to hold the heat absorbed.

When sufficiently dry, the meat pieces are either sold without prior packaging or wrapped in jute sacks. Plastic sacks are not suitable, because the product still contains a certain proportion of its original moisture content, and this moisture must be allowed to drain freely from the product. Charque keeps for months under ambient room conditions and is resistant to infestation by insects and growth of moulds.

Meat Preservation by Thermal Treatment

The prolonged shelf-life of heat-treated meat and meat products is achieved through reducing growth of, or inactivating, micro-organisms by a thermal process. The principal steps of the heat preservation method are to:

- place the product in a container (can, glass jar, pouches of synthetic material or laminate with aluminium) which is hermetically sealed after filling and which is impermeable to any external substances; and
- submit the hermetically sealed product to thermal treatment with a defined temperature and time combination.

Equipment for Thermal Treatment

Thermal or heat treatment is done by submerging the products in cooking vats or pressure cookers which contain hot water or steam or a mixture of both. It can be performed under pressure in pressure cookers (retorts, autoclaves) in order to reach temperatures above 100°C ("sterilization"). Sterilization is the most important and efficient type of heat treatment, since foods free from viable micro-organisms can be obtained and most of these products can then be stored without refrigeration. In contrast, temperatures up to 100°C can be achieved in simple cooking vats ("pasteurization"). A certain amount of micro-organisms resist this moderate heat treatment and the resulting pasteurized products must consequently be stored under controlled temperatures.

In simple retort cookers (autoclaves) pressure is generated either by a direct steam injection, by heating water up to temperatures over 100° C or by combined steam and water heating. The retort must be fitted with a thermometer, a pressure gauge and a relief valve. Modern autoclaves may also have revolving drums, speeding up the heating of the products.

After thermal treatment the product must be chilled as quickly as possible, in order to avoid overcooking. Hence, this operation is done within the cooker by introducing cold water. The contact of cold water with steam causes the latter to condense with a rapid pressure drop in the retort. However, overpressure simultaneously built up during thermal treatment within the cans, jars or pouches remains for a certain period, and may induce permanent deformation or damage of these containers. Therefore, a high pressure difference between the cooker and the internal pressure in the containers must be avoided. This is generally achieved by a blast of compressed air into the retort or by sufficient hydrostatic pressure of the introduced chilling water.

Containers for Thermally Treated Preserves

Containers for heat-preserved food must be airtight in order to avoid recontamination by environmental microflora. Moreover, no traces of undesirable substances which the packaging material may contain, such as heavy metals (lead, tin), should be permitted to migrate into the product. Currently, most of the thermally preserved products are in metal containers (cans), others are packed in glass jars or plastic or aluminium/plastic laminated pouches.

Metal containers are *cans* (tins) produced from *tinplate* or *tin-free* steel. They are usually cylindrical. However other shapes such as rectangular or pear-shaped cans are also encountered. Tinplate consists of steel plate, plated with tin on both sides. The steel body usually is 0.22 to 0.28 mm in thickness. The tin layer is from 0.385 to 3.08 ìm. Tin-free steel plates have other protective coatings such as chromium, aluminium, or nickel, which are generally even thinner than the tin layers of the tinplate.

The cans (tins) usually consist of three elements, i.e. the body and two ends. The seaming panel of the ends is fitted with a synthetic lining. The ends are fastened to the body with seams made by a seaming (closing) machine. Proper seaming is vital to the tightness of the can. Any leak causes recontamination, in particular during chilling. This will result in swelling of the cans during storage, and creates a risk of food poisoning.

For smaller and easy-to-open cans aluminium is frequently used. *Aluminium cans* are deep-drawn, i.e. the body and the bottom end are formed out of one piece and only the top end is seamed on after the filling operation. The advantages of aluminium cans are low weight, resistance to corrosion, good thermal conductivity and recyclability, but these cans cannot be soldered or welded. They are less rigid and more expensive than steel plate.

Glass jars are used less often for meat products because of their fragility. They consist of a glass body, and a metal lid. In households, glass jars with glass lids are often used. The seaming panel of the metal lid has a lining of synthetic material. Glass lids are fitted by means of a rubber ring.

Containers made either of *synthetic material* or *laminates of aluminium foil with synthetic material* are of growing importance in thermal preservation. Heat-resistant plastic pouches, which are closed by clip, are usually made of polyester (PETP) and used for frankfurters in brine or ready-to-eat dishes. From laminated films, for instance, polyester/polyethylene (PETP/ PE) or polyamide/polyethylene (PA/PE), relatively rigid containers can be made, usually by deep drawing, which are used for filling with pieces of cured ham or other kinds of prepared meat. Widely used for small portions, particularly of sausage mix, are round containers formed out of a laminate of aluminium foil and polyethylene (PE) or polypropylene (PP). PE or PP permit the heat-sealing of these containers, which can then even be subjected to intensive heat treatment.

Meat Products Suitable for Canning

Basically all meat products which require heat treatment to prepare them for consumption are also suitable for heat preservation. Only meat products which do not receive any form of heat treatment before being consumed, such as dried meat, raw hams or dry sausages, are naturally not suitable for canning. These products are conserved by a low pH value and/or low water activity.

The following groups of meat products, when not consumed freshly cooked, are frequently found as canned products:

- cooked ham
- sausages with brine of the frankfurter type
- sausage mix of the bologna or liver sausage type
- meat preparations such as corned beef, chopped pork, etc.
- ready-to-eat dishes with meat ingredients such as beef in gravy, chicken with rice, etc.
- soups with meat ingredients such as chicken soup, oxtail soup, etc.

Organoleptic, Physical and Microbiological Aspects of Thermal Treatment

The intensity of heat treatment has not only a decisive impact on the inactivation of micro-organisms, but also on the organoleptic quality of the product. There are products which undergo intensive temperature treatment without significant losses in quality. On the other hand, other products may deteriorate considerably in taste and consistency after sterilization. In these cases less intensive thermal treatment is required but, at the same time, other hurdles, such as low pH value and/or water activity or a lower storage temperature, have to be built up in order to inhibit bacterial growth.

The intensity of thermal treatment can be defined in physical terms. The term widely used under practical conditions is the F-value, with which the lethal effect of heat on micro-organisms can be defined. The thermal death time for different micro-organisms calculated at 121°C and expressed in minutes, is used as the reference value.

The thermal death time for spores of *Clostridium botulinum* at 121°C is 2.45 minutes or in other words, an F-value of 2.45 is needed to inactivate all these spores in the product at 121°C. Spores of other micro-organisms are more or less heat resistant. Vegetative cells of micro-organisms are generally destroyed at temperatures of less than 100°C and therefore play no role in the F-value calculations. The definition of the F-value at 121°C is as follows:

— F = 1: lethal effect at 121°C on micro-organisms after 1 minute

— F = 2 (3, 4, etc.): lethal effect at 121°C on micro-organisms after 2(3, 4, etc.) minutes. In Tables 2 and 3 some examples are given for F-values obtained at different time/temperature combinations:

Table 2. F-values corresponding to various temperatures

95°C	per minute:	F = 0.003
100°C	per minute:	F = 0.008
105°C	per minute:	F = 0.025
110°C	per minute:	F = 0.079
115°C	per minute:	F = 0.251
121°C	per minute:	F = 1.0
125°C	per minute:	F = 2.51
130°C	per minute:	F = 7.94

The lethal effect can be shown in the reduction (in percentage) of the total number of micro-organisms present in the product. The destruction of micro-organisms is at an exponential rate, which means that the higher the initial bacterial load (using the same time-temperature combination), the higher the number of surviving bacteria.

Highly contaminated raw material with bacterial loads of 10 million per g will, even after intensive heat treatment, still give final products with a rather limited shelf-life because of the high remaining rate of contamination.Since the heat treatments will in many cases not be intense enough to destroy all spores, it is important that cans be chilled as rapidly as possible after retorting and that storage temperatures generally not exceed 20 to 25°C.

The nature of the heat-preserved product, its pH, amount of salt and other curing agents, and the number of spores present, together with retorting time and temperature, determine the degree of commercial sterility and product safety. It has been shown that F-values of 4 in heat-preserved products will guarantee commercial sterility. Products with F-values below this level need additional measures such as lowering the pH or the a_w or refrigerated storage for their microbiological safety.

Micro-organisms have two adverse effects in improperly treated heat-preserved products:

- organoleptic deterioration through protein degradation;
- food poisoning by bacteria and/or toxins.

The food-poisoning aspects require special care during production and storage of heat-treated preserves, bearing in mind that some heat-resistant micro-organisms are able to produce dangerous toxins, amongst them *Clostridium botulinum*, which may have fatal consequences.

Practical Application of F-values

By measuring the temperature of the product periodically during thermal treatment, the final F-value can be determined. It is obvious that during thermal treatment the product temperature will rise constantly. The temperature taken in the centre of the container after each minute of heat treatment corresponds to a certain F-value. These partial F-values are added up (for example by using special tables containing F-values corresponding to temperatures from 90°C to 140°C) and the sum is the overall F-value of the product.

The exact F-value is of special importance for the producer because:

- it ensures appropriate thermal treatment of the product, thus avoiding over-or undercooking;
- it enables the product's storage time to be determined.

In practice it is not necessary to calculate the F-value repeatedly for the same type of batch processed in the cannery. The F-value can be determined once for each batch according to the size of the containers and intensity and duration of thermal treatment. If these parameters remain unchanged, the F-values will not be subject to alteration.

Categories of Heat-treated Preserves

Pasteurized products

Only slight thermal treatment. Temperatures reached in the product centre are in the range of 82°C and below 100°C ("pasteurization"). The F-value cannot be determined, remaining almost at zero.

— Inactivated: most vegetative micro-organisms
— Not inactivated: spores of *Bacillus* and *Clostridium*
— Storage required: uninterrupted cold chain (2–4°C), up to six months

Cooked preserves

Thermal treatment only with boiling water (no pressure cooker).

Temperature reached in the product centre is up to 100°C. Low F-value.

— Inactivated: all vegetative micro-organisms
— Not inactivated: spores of *Bacillus* and *Clostridium*
— Storage required: not higher than 10°C for one year. Spores will not grow under these conditions.

"Three-quarter" preserves

Thermal treatment in pressure cooker. Temperatures reached in the product centre are between 108 and 112°C. F-value 0.6 to 0.8.

— Inactivated: all vegetative micro-organisms, spores of *Bacillus*
— Not inactivated: spores of *Clostridium*
— Storage required: not higher than 15°C for one year. Spores of *Clostridium* will not grow under these conditions.

"Full" preserves stable under temperate conditions

Intensive thermal treatment in pressure cooker. Temperature reached in the product centre is about 121°C. F-value 4 to 6 ("sterilized product").

— Inactivated: all micro-organisms except thermophilic spores
— Storage required: ambient temperature (for one year), but not tropical conditions (40°C or more).

"Full" preserves stable under tropical conditions

Very intensive thermal treatment, with a long period of 121°C or higher in the product centre. F-value of 12 and more.

- — Inactivated: all micro-organisms including thermophilic spores
- — Storage required: ambient temperature even under tropical conditions (up to four years).

Shelf-stable preserves

This group of preserves is different from those mentioned previously, since preservation is achieved not only by thermal treatment, but also by utilizing other means to prevent microbiological growth such as nitrite, low water activity and/or low pH. This combined effect has the advantage of a fully shelf-stable product under all ambiental conditions without undergoing intensive thermal treatment (less than 100°C) and without major losses in organoleptic quality.

Shelf-stable preserves are a fairly new development in the food sector and will certainly gain special importance in countries without an uninterrupted cold chain.

Thermal treatment of different intensity for different products is used to avoid deterioration, which varies from product to product. The following table gives some examples on how thermal treatment should be conducted. It is a general rule in this context that products in smaller containers can undergo more intensive thermal treatment because of faster heat penetration.

Impact of Packaging Methods on Meat Preservation

In most developing countries meat carcasses are handled and stored either without or with only minimal refrigeration. A fast-turnover system is used, ensuring that meat slaughtered during the night and morning is sold and consumed the same day. This well-proven traditional method functions satisfactorily because in the short period between slaughtering and consumption, micro-organisms cannot increase to the extent of spoiling the meat and making it inedible. This system is eminently suitable for rural villages and small towns in many developing countries, particularly because local cooking habits do not call for ripened and tender meat.

This fast turnover of meat normally does not lead to meat spoilage, although massive contamination during slaughtering and meat handling frequently occurs. Fortunately, in most cases toxic micro-organisms do not develop to such an extent to threaten consumers' health, provided that the time between slaughter and meat cooking is short and cooking is done properly and with sufficiently high temperatures.

In traditional meat handling, fresh meat is generally not packaged at all or just wrapped in paper, leaves, etc. Meat, traditionally preserved by drying, is sometimes packaged in linen bags, baskets or pottery to facilitate storage and transport and to provide some kind of protection against dirt, insects, etc.

With higher concentrations of population, however, this traditional system now becomes outmoded in some places in developing countries because more time is needed between slaughtering and ultimate consumption. Meat frequently has to be stored, transported, prepared and distributed through a retailer or supermarket, all of which is considerably time-consuming. In order to safeguard fresh meat during this extended time, certain methods of preservation have to be applied. Refrigeration is the obvious solution, but this is expensive and therefore frequently not available in developing countries. Energy-saving storage methods are therefore particularly relevant in underdeveloped areas. For both methods, either using the refrigeration technology or energy-saving methods to extend the shelf-life of meat and meat products, proper packaging has an important part to play.

Purpose of Packaging Fresh and Processed Meat

The purpose of packaging is primarily to protect foodstuffs during the distribution process, including storage and transport, from contamination by dirt, micro-organisms, moulds, yeasts, parasites, toxic substances or those influences affecting smell and taste or causing loss of moisture. Packaging should help to prevent spoilage, weight losses and enhance customer acceptability. Simple packaging without further treatments is less effective in prolonging the shelf-life of meat and meat products. Frequently full advantage of packaging can only be achieved in combination with preservation methods.

Factors Affecting the Shelf-life of Meat and Meat Products

Though meat handling, storage and consumption may differ from one place to another, the factors limiting the shelf-life of these products are the same.

There are *endogenous* factors, such as:

- pH-value or the degree of acidity of the product;
- a_w value or the amount of moisture available in the product; and *exogenous* factors, such as:

- oxygen (from the air);
- micro-organisms;
- temperature;
- light; and
- evaporation and desiccation.

Combined Effect of Ph- and A_w-value

Generally speaking the shelf-life of meat and meat products will be longer the lower the pH-value and/or a_w-value. Both factors (either pH or a_w alone or the two together) have a decisive influence on the growth of micro-organisms in food. However, there are limits for most meat products regarding decreased pH-value and a_w-value, particularly for organoleptic reasons. Except for some special products, consumers do not want meat products to be excessively acidic or dry.

Uncanned meat products can be classified into three storage groups according to their pH and a_w. Each group requires different storage conditions.

Highly perishable meat products have a pH-value above 5.2 and an a_w-value above 0.95; refrigeration at or below +5°C is needed. These are raw fresh meat (without additives), bologna-type sausages, cooked sausages and cooked ham.

Perishable meat products have a pH-value below 5.2 or an a_w-value below 0.95. Refrigeration at or below +10°C is needed to keep them stable. Products such as meat or poultry pieces in vinegar jelly (acid) and semi-dry sausages or hams belong to this group.

Shelf-stable products have a pH-value of or below 5.2 and an a_w-value of or below 0.95, or only a pH-value below 5.0, or only an a_w-value below 0.91. No refrigeration is required in these cases, the products remaining stable under ambient temperatures. The most common products in this group are the various kinds of dried meat.

Under the above conditions no microbial growth in meat and meat products will occur. However, this does not mean that the products remain stable for an undetermined period. Their shelf-life will be limited by chemical or physical deterioration, by rancidity and discoloration. In this situation the product quality will benefit from the application of suitable packaging materials, which reduce the physical and chemical influences on the product or protect the product completely. The following noxious influences may occur.

Oxygen

The oxygen content in the air is about 20 percent. If oxygen affects meat and meat products during prolonged storage periods, it will change the red colour into grey or green and cause oxidation and rancidity of fats with undesirable off-flavours.

The foils used for food packaging differ in their permeability to oxygen. The lower the oxygen permeability of the packaging material, the more efficient will be the protection of product quality. The best protection will be achieved using oxygen-proof packaging films together with vacuum packaging of the product. This ensures that practically no oxygen is left in the package and no oxygen will penetrate from the air into the product.

Light

The prolonged exposure of meat and meat products to daylight or artificial light accelerates oxidation and rancidity because light provides the energy for these processes.

Transparent packaging films give no protection against light influences. Therefore, for products under strong light exposure, coloured or opaque films should be preferred. Films laminated with aluminium foil are absolutely impermeable to light. Products in transparent packaging film are sufficiently protected when kept in the dark or under moderate illumination.

Evaporation

Fresh foods with a relatively high moisture content such as meat, fresh sausages, cooked ham, etc. will have considerable losses of weight and quality by evaporation during storage if they are not packed. The packaging material must therefore be sufficiently vapour-proof. Most plastic films used for food packaging comply with this requirement.

Secondary Contamination

During slaughtering, carcass dressing, meat cutting and/or processing, the contamination of meat to some extent cannot be avoided. The further growth of micro-organisms in meat and meat products cannot be stopped through packaging only. However, secondary contamination of these foods, for example by contact with dust, dirty surfaces and hands, can definitely be prevented through proper packaging, preferably with plastic films which are absolutely impervious to agents causing secondary contamination.

Suitable Materials and Equipment for Packaging Meat and Meat Products

Packaging films can be subdivided into *cellulose films, plastic films* and *aluminium foil.* They can either be used as monofilms or as two or more different films laminated together. These materials differ in:

- oxygen permeability;
- water vapour barrier;
- resistance to hot and cold temperatures; and
- mechanical strength.

Nearly all available films are of thermoplastic materials, and therefore *heat sealable*, resulting in hermetically sealed plastic pouches, bags, etc.

A high oxygen barrier is important in the application of films for packaging meat and meat products. Films made of polyvinylchloride (PVC), polyethylene (PE) or polypropylene (PP) have a relatively high oxygen permeability, whereas polyvinylidenchloride (PVDC), polyester (PETP), polyamide (PA) and cellulose film (ZG) are less or almost non-permeable to oxygen. The latter materials are therefore better suited for packaging meat and meat products. However, the materials of the first group are frequently used as laminates with materials of the second group in order to achieve special effects regarding mechanical strength, heat sealing properties or making the package practically impermeable to both oxygen and water vapour.

For the efficient utilization of these materials, air must be drawn out completely from the packages with the meat product ("vacuum-packaging") and the package should be hermetically closed by heat sealing or metal clip. Vacuum-packaging will inhibit the growth of yeasts, moulds, and most aerobic bacteria under refrigeration temperatures. However, facultative anaerobes such as acid-producing bacteria grow freely under adequate a_w-conditions. These micro-organisms do not threaten the health of consumers but may affect the quality of the products by negatively influencing colour and taste. Therefore efforts should be made to keep the initial number of bacteria low, which can be obtained by simple hygienic measures during cutting and packaging, such as frequent handwashing, clean tools and clean tables.

The above general considerations refer to vacuum-packaged meat and meat products with a high moisture content. All these products are shelf-stable because of their low water activity and therefore do not require refrigeration. They are partly dried in the sun, or others are dried and smoked over a wood fire.

These low-moisture products are hygroscopic and should be protected in the case of high air humidity. Traditional wrapping methods using paper, jute or linen will not inhibit the penetration of water vapour. However, a simple polyethylene bag is sufficient as a water vapour barrier, but care must be taken that the bag is tightly closed. If plastic bags are used for packing, then the pieces of dry meat should be cut to a certain length so that they can be better arranged in the bags. The bag should be bound with rope, rubber or cellotape. Bags stuffed with meat in this way are best stacked into a carton.

Longer storage times for traditionally preserved dried meat can be achieved with vacuum-packaging in laminated films, which protect against humidity and oxidation by oxygen from the atmosphere. Vacuum-packaging can also gain importance for traditionally preserved products. Although they are shelf-stable, exposure of these products to oxygen and the humidity of the surrounding atmosphere will cause unfavourable alterations (rancidity, loss of flavour, growth of certain micro-organisms) over the long term. Air and vapour-tight packaging will ensure a prolonged shelf-life.

Simple vacuum-packaging machines, suited to less developed regions, are available. They consist of a chamber with a removable lid, so that the plastic pouch or bag containing the product can be placed inside the chamber, which is evacuated with an electrically driven vacuum pump. By closing the lid, air can be drawn out of the chamber and also out of the open bag. While still in the vacuum chamber, the bag is hermetically closed by heat sealing by means of a manual or automatic electrically heated sealing device. After being sealed the bag can be exposed to ambiental conditions without air penetrating the package.

Basic Methods of Quality Control

The quality of meat and meat products is defined by the following criteria:

- palatability (typical texture and consistency, juiciness, good flavour);
- proportion of lean meat to fat;

- freshness and adequate conservability of the products;
- absence of harmful micro-organisms or substances; and
- appropriate (preferably minimal) use of additives and meat extenders.

The different criteria need different methods of quality control, such as:

- organoleptic evaluation
- physical test methods
- chemical analysis
- microbiological examination

According to the accuracy needed, the control method applied can be simple or more complicated and different auxiliary technical devices must be used.

In order to inform consumers and meat processors about the quality of meat and meat products, simple and fast control methods are best suited in many cases, although exact details on residues, toxins and special food components can only be obtained through specialized laboratories.

Basic methods for quality control must involve little or no equipment and obviously sensory evaluation will be most important. Some physical tests, however, can easily be performed using simple instruments such as thermometers, manometers, scales, etc.

By contrast, chemical and microbiological tests are more complicated. These methods not only require standard equipment but also skilled and experienced personnel to do the tests and to interpret the results.

The following mainly refers to the basic methods of quality control used in connection with the handling and processing of meat. These control methods can easily be applied for meat products processed with simple meat preservation techniques.

Organoleptic Evaluation

Organoleptic evaluation consists in describing the attributes of food, in this special case of meat and meat products, that can be perceived by the sense organs. The attributes to be evaluated are appearance, colour, texture and consistency, smell and taste.

Appearance

The way meat looks, either as a carcass or as boneless meat cuts, has an important impact on its objective or subjective evaluation. Grading is an objective evaluation method in this context. Traditional methods of carcass grading after slaughter involve the aspect of beef or pork sides, poultry carcasses, etc. Skilled graders are able to classify different carcasses by checking the size, the volume of muscular tissue, fat layers, etc. Although in modern grading procedures more and more technical equipment has been incorporated, visual methods are still in use. They can be of special value in most developing countries where no extremely sophisticated methods are needed.

The way the consumers or the processors check the appearance of meat is subjective. Differences will be registered in the relation of lean meat and fat including the degree of marbling or in the relation of bones and lean meat. Furthermore, unfavourable influences can be detected such as unclean meat surfaces, surfaces too wet or too dry, or unattractive blood splashes on muscle tissue.

Processed meat, on the other hand, can roughly be evaluated by its appearance according to the different raw materials of which the product is composed and where the use of some components is exaggerated (for instance too many particles of visible fat or connective tissue, etc.). Special product treatments (for instance chilling, freezing, cooking, curing, smoking, drying) or the kind and quality of portioning and packaging (casings, plastic bags, cans) will be recognized by evaluating the appearance.

Colour

Under normal circumstances the colour of meat is in the range of red and may differ from dark red, bright red to slightly red; but also pink, grey and brown colours may occur. In many cases the colour indicates the type and stage of the treatment to which the meat has been subjected, as well as the stage of freshness.

In judging meat colour, some experience is needed to be able to distinguish between the colour which is typical for a specific treatment or which is typical for specific freshness. Furthermore, meat deriving from different species of animals may have rather different colours, as can easily be seen when comparing beef, pork and poultry meat.

The natural colour of fresh meat, except poultry meat, is dark red, caused by the muscle pigment, myoglobine. Fresh meat surfaces which have been in contact with the air for only a short period turn into a bright red colour because of the influence of the oxygen in the air. Oxygen is easily aggregated to the myoglobine and drastically changes the colour of the meat surfaces exposed to it. On the other hand, in the absence of oxygen, for example in meat cuts packaged in impermeable plastic bags, meat surfaces remain or become dark red again. The same conditions generally prevail in the interior of meat cuts which are not reached by oxygen. Changes from dark red to bright red are therefore typical and are normal reactions of fresh meat.

Meat which is in the process of losing its freshness, however, no longer shows a bright red colour, even when intensively exposed to the air, because of the partial destruction of the red meat pigment which results in a grey, brown or greenish colour. Once these conditions occur the consumer has to decide, after carefully checking the appearance, together with testing smell and taste, whether the meat has to be discarded as a whole or whether use can be made of some parts which so far have not been altered.

Remarkable changes in the meat colour occur when fresh meat has been boiled or cooked. It loses its red colour almost entirely and turns to grey or brown. The reason for this is the destruction of the myoglobine through heat treatment. On the other hand, it has long been known that after pickling (curing) fresh meat with curing ingredients (nitrite), the meat colour remains red during longer storage periods, after ripening, drying and even after intensive heat treatment. Obviously the original meat colour has not been conserved, but a chemical reaction has taken place during the curing process transforming the unstable pigment of the fresh meat into a stable red pigment. This is the typical colour shown in sausages of all types, raw and cooked hams, corned beef, etc.

It should also be noted that cured products have a longer shelf-life than fresh meat because of the conserving effect of the curing salt. However, cured products will also deteriorate under unfavourable conditions, cooked cured products sooner than raw cured products. Cured products with a decreasing keeping quality can be recognized when the red colour becomes pale or changes to grey or green.

Texture and consistency (tenderness and juiciness)

Meat prepared for the consumer should be tender and juicy. Meat tenderness depends on the animal species from which the meat originates. Lamb, pork and poultry meat are sufficiently tender after slaughter, but beef requires a certain period of maturation to achieve optimal eating quality.

Texture and consistency, including juiciness, are an important criterion, still neglected by many consumers, for the eating quality of meat. Often consumers do not know that the eating quality of meat can be upgraded by ripening, especially in the case of beef and similar meats. There is also a great deal of consumer negligence in how to prepare meat. It should be cooked to become sufficiently tender, but cooking should not be too intense otherwise the meat becomes dry, hard and with no juiciness.

The simple way to check the consistency of foods is by chewing. Although this test seems easy, in practice it is rather complicated. Taste panelists need experience, particularly when the different samples have to be ranked, for example which sample is the toughest, the second toughest or the most tender.

The texture is of less importance in meat products, such as cured or canned products, sausages, etc., because they are either made of comminuted meat and/or meat which has undergone heat treatment or long maturation periods and will therefore generally be tender. On the other hand, inadequate processing methods (too intensive cooking, curing, comminuting) may cause losses in the desired consistency and juiciness, and the best way to check this is by chewing.

Smell and taste (aroma and flavour)

These characteristics are related to each other to a certain extent because they have to be evaluated together for the reliable determination of a product's flavour. The smell of fresh meat should be slightly acidic, increasing in relation to the duration of the ripening period because of the formation of acids such as lactic acid. On the other hand, meat in decomposition generates an increasingly unpleasant odour owing to substances originating from the bacterial degradation of the meat proteins, such as sulphur compounds, mercaptane, etc.

The freshness of meat is generally indicated by its smell together with its appearance and colour. Sorting out deteriorated meat is mandatory from the point of view of the product's palatability. It is also important because of the fact that high bacterial contamination of meat in decomposition could be accompanied by food-poisoning bacteria(pathogens), which have a deleterious impact on consumers' health. On the other hand, the best fresh meat can also be heavily contaminated with food-poisoning bacteria because these micro-organisms do not cause organoleptic alterations by destruction of meat proteins. Food poisoning can therefore only be avoided by proper hygienic meat handling. The flavour of fresh meat can also be checked by putting small samples (approx. 10 pieces of 1 cm^3each) in preheated water of 80°C for about five minutes (boiling test). The odour of the cooking broth and the taste of the warm meat samples will indicate whether the meat was fresh or in deterioration or subject to undesired influences, for instance rancidity of the meat fat, any a typical meat flavour due to the feed and the sex (boar taint) of the animal or treatment with veterinary drugs shortly before slaughter.

When processing the meat, the smell and taste of the meat products can differ a great deal owing to heat treatment and the use of salt, spices and food additives. Every meat product has its typical smell and taste, and the test person should know about it. Changes in these qualities indicate the use of improper raw materials or a deterioration of the meat product during storage.

Experience is required to become acquainted with the typical flavour (smell and taste) of foods. Only four basic taste components—sweet, sour, bitter and salty—will be perceived by the taste buds. These receptors are small papillae located in certain areas of the tongue. However, the overall flavour consists of smell and taste produced by the meat components and influenced and covered by spices and those compounds produced by ripening or heat treatment. Flavour test panelists should be aware of these special cases. Panelists should not smoke or eat spicy meals before starting the test and should rinse their mouth frequently with warm water during the test.

Sensory evaluation plays an important role in the examination of meat and meat products. Not only does scientific sensory evaluation with skilled panelists using special test programmes and point systems give reliable results, but useful results can also be obtained in a simple way at the consumer level. For the average consumer sensory evaluation is the only way to decide whether or not he or she should buy or eat a certain product.

In developing countries consumers do not receive sufficient information and training on this point, although it is often the only means available for quality control. Sensory evaluation is easy to understand and to perform. What is needed is a basic knowledge of the composition of foods and their typical texture, colour and flavour.

Physical Test Methods

Physical test methods focus either on the actual condition of meat and meat products, or on the conditions around the product, for example in storage rooms, packages, etc. Equipment will be needed for all these tests which is easily applicable and resistant to utilization under the conditions of practical meat handling and processing.

Temperature

Storage of meat and meat products requires low temperatures to make sure that the growth of micro-organisms will be retarded (chilling between-1 to +4°C) or inhibited (freezing preferably in the range of -18 to -30°C).

Cooking of meat requires high temperatures (starting from a temperature of about 55°C needed for denaturation, but generally higher temperatures are applied, up to 100°C).

Canning of meat requires temperatures above 100°C, and for sterilized products where all micro-organisms are inactivated, at least 121°C.

These examples demonstrate the importance of different temperatures for different purposes and the necessity of exact temperature measurements using thermometers or temperature recorders.

Glass thermometers should not be used in direct contact with meat because they may break, leaving undesired fragments in the meat, but they are useful when permanently fixed to walls of chillers or production rooms or to cooking equipment or autoclaves for easy checking of the relevant temperatures.

Mechanical bimetal-thermometers, utilizing the extension or contraction of a bimetal spiral under various temperatures, are not very accurate and not sufficiently shock-resistant for practical work in meat industries. Nevertheless, they are widely used and can serve for rough estimates.

Electrical thermometers, consisting of a sensor and a battery-powered electronic instrument, are well suited for meat industries. The sensor functions as a semiconductor. Under different temperatures, differences in the electric conductibility of the sensor are produced. The temperature which the sensor takes up by contact to the surrounding media (water, air, meat, etc.) produces a certain voltage in the electric system. This voltage is registered and displayed as a digital reading of the actual temperature on the instrument.

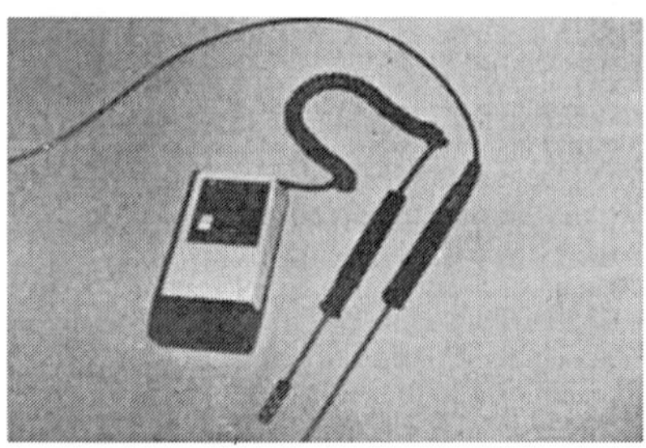

Figure 4. Electrical thermometer with digital display and two sensors for measuring air temperature (left) and the temperature of meat, liquids, etc. (right).

Advantages of modern electronic thermometers are:

- no glass or other parts that can easily break;
- the sensor can be easily pushed deep into the meat, as well as into frozen meat, and is also heat-resistant under sterilization temperatures;
- display of the correct temperature within seconds;
- no frequent calibration necessary;
- a wide range of temperatures can be covered using one instrument (the temperature range of the instruments recommended for use in meat industries should be between +140°C and -40°C); and
- accurate temperature measurement, also in decimals.

Humidity

In some special field of meat processing and storage, air humidity is of importance.

In cutting rooms the humidity of the air should be below the level which would cause vapour condensation on the surfaces of the meat being deboned and cut. Vapour condensation may enhance bacterial growth.

Storage chillers for fresh meat require a balanced air humidity that does not cause wet surfaces on the meat with resulting accelerated bacterial growth, but on the other hand keeps evaporation losses low. The relative humidity recommended for this special purpose lies in the range of 70 percent.

Chambers for the maturation of raw hams or dry sausages of the salami type require controlled air humidity, starting from 90–95 percent and after a certain period finalizing the process at 70–75 percent relative humidity. This procedure is important for the balanced drying and ripening of the products. Suitable instruments (hygrometers) for the exact measurement of relative humidity are therefore needed.

In simple but less accurate hygrometers a hair or special synthetic fibre is connected with a pointer and, according to the contraction of the hair or fibre under dry conditions and its extension under wet conditions, the pointer indicates the actual relative humidity on an appropriate scale.

A more accurate way for humidity control is the psychrometric system. These instruments use a dry and a wet sensor to define the ambiental temperature. The temperature indicated by the wet sensor will always be lower, in this case, because of evaporative cooling. The drier the air, the more intensive evaporative cooling will be. Using both temperature values (dry and wet temperature), the value of the relative humidity is determined in practical work using a table for easy calculation.

A modernized psychrometric system which uses electronic devices is available. In this case the humid sensor has not actually to be kept wet, but consists of hygroscopic material with altering electrical resistance. The relative humidity calculated from the temperatures delivered by the sensors by means of a micro-processor is directly displayed on the instrument.

Water activity (a_w)

Water activity is the free water available for microbial growth in a food product. Free water is that part of the water content that can be eliminated from the product in the form of water vapour. Hence, the term "water activity" is defined as the ratio of the water vapour pressure measured in the product and the pressure of a saturated water vapour atmosphere at the same temperature. This physical definition is used in connection with a number of meat products whose keeping quality depends on their water content. Micro-organisms need a certain degree of moisture to be able to grow on foods. The minimum moisture content necessary for microbial growth varies with the single species of micro-organisms and can be expressed in terms of minimum water activity required.

The lowest a_w-values permitting growth of spoilage organisms are:

- normal bacteria 0.91
- normal yeasts 0.88

- normal moulds 0.80
- halophilic (NaCl-tolerant) bacteria 0.77.

The keeping quality of dried meats and meat products without refrigeration depends on their water activity. Dried meat such as biltong, charque, etc. reaches a sufficiently low water activity to be shelf-stable. However, water activity should decrease as fast as possible as slow drying could cause deterioration in a prolonged phase of the process with a still high water activity. The situation is more complicated in the case of products which cannot be dried too intensively such as dry sausages or raw hams. The water activity of these products is relatively low but would still allow the growth of some undesired micro-organisms. Under these circumstances an appropriate shelf-life has to be ensured by the combination of several inhibiting factors, i.e. water activity, content of salt and curing ingredients and the acidity of the product.

Information on the water activity of certain products can be important for further handling, packaging and storage. Simple methods for the determination of water activity are therefore useful.

As water activity refers only to the water available for microbial growth in a product, the chemical analysis of the total moisture content is of limited value since it would also include the water bound by the proteins. The proper way to determine water activity is to measure the humidity of the remaining air in a hermetically closed small cabinet which is to a certain extent filled with the product sample. After a short time a hygroscopic equilibrium between the sample and the surrounding air will be reached. Thus, the humidity determined in the air is equivalent to the humidity available in the product and water activity can be calculated.

For the measurement of air humidity under these conditions, the same principles apply as previously described. Simple devices utilize the extension or contraction of hairs or synthetic fibres, and more sophisticated and more expensive systems use electronic devices.

The sample is placed in the bottom part of the tin and then the lid of the tin that contains the device to measure the humidity is hermetically screwed on. After two hours, hygroscopic equilibrium is reached in the can and the reading of the instrument corresponds to the actual water activity of the product, provided the test has been carried out at a temperature of exactly 25°C. If this temperature cannot be maintained, corrective calculations will be necessary.

Some examples for values of water activity (a_w) of different products are shown:

fresh raw meat	0.99–0.98
cooked ham	0.98–0.96
frankfurter-type sausages	0.98–0.93
liver sausage	0.97–0.95
raw cured ham	0.96–0.80
dry sausage (salami type)	0.96–0.70
dry meat	0.75–0.50

A certain number of micro-organisms are inhibited at a_w 0.95, but other species are still able to grow. At a_w 0.92 all bacteria groups are inhibited, but the growth of moulds and yeasts is still possible.

Airtight closure of cans

For shelf-stable canned meat products two aspects are important from the microbiological standpoint. During sterilization, micro-organisms and their spores have to be inactivated and the can must be hermetically sealed to avoid further contamination of the product after sterilization.

Invisible small perforations of the tinplate or small defects after the closure of the lid will inevitably lead to a *recontamination* by penetrating bacteria and after a short time spoilage of the canned product will occur. Cans should therefore be checked from time to time for these defects.

A simple method is available for this purpose. Using an air-pump with a special device to penetrate the tinplate, the air is pumped into a closed but empty can. The internal pressure built up in the can can be controlled by a manometer connected with the air pump. When dipping the inflated can into water it can easily be seen whether the can is hermetically sealed and if not where the cause for the permeability is, either in the prefabricated body (side wall and bottom) of the can or in the area of the lid seam. In the latter case the function of the can-closing equipment in the processing plant has to be thoroughly checked.

Weight differences

The high water content of meat (approx. 70 percent) and meat products (which varies from 70 percent to 10 percent) causes weight differences owing to evaporation losses or drip losses that occur during handling, processing or storage.

Unpackaged meat and meat products are especially subject to considerable evaporation losses. During chilling of warm carcasses evaporation losses of 1 to 2 percent cannot be avoided but further evaporation losses of chilled or frozen meat should not occur when suitable storage conditions (not too dry) or suitable packaging (plastic bags, containers, boxes) are employed. However, some drip losses of packaged meat cannot be avoided.

During meat processing weight losses of meat by cooking, frying or other heat treatment can be registered and reach high values (up to 30 to 35 percent). These losses are unavoidable.

On the other hand, some meat products require weight losses by evaporation to reach their specific keeping quality, for example raw hams, dry sausages or dried meat. In this case, water activity as previously described plays an important role.

Information on weight losses in meat handling and processing is important from the economic and technological point of view. Weight losses can easily be determined using scales of different types, such as suspension scales for carcasses or batches of products and horizontal scales for packages or portions.

Salt concentration in brines

In addition to dry curing methods (dry salt and curing ingredients on the meat), brines are also used for pickling and curing the meat. Brines contain salt and in most cases also sugar and nitrite dissolved in water. With this curing process, meat is either immersed into a brine or the brine is injected into the meat with special devices. In both cases salt is a limiting factor for the sensory quality of the products; in other words salt is needed but should not exceed 2.5 to 3 percent in cooked cured products and 4.5 to 5 percent in raw cured and dried products.

To comply with these requirements, simple methods for testing salt concentration in brines are necessary. For this purpose salimeters have proved to be a useful piece of equipment. Salimeters are densimeters, the graduation showing salt concentrations. Salimeters are dipped into the brine and according to a lower or higher salt content they sink deeper or less deep into the brine. The reading of the salimeters at surface level indicates the salt concentration of the brine. The various technologies of meat curing use brines with NaCI-concentrations in the range of 8 to 22 percent.

Other Physical Test Methods

The physical test methods which have been described can easily be performed since the use of the technical equipment necessary is not too complicated. Other physical test methods exist too, for example, light intensity measurement, colour measurement or texture and consistency measurements of meat and meat products. These tests require rather complicated and expensive instruments and skilled technical personnel. For routine work, criteria such as light, colour, texture and consistency can be evaluated in a satisfactory way by using the corresponding sensory test methods.

Chemical analysis

Chemical characteristics of foods are related to the product itself and refer primarily to the content of specific substances, which are important from the point of view of keeping quality, flavour, nutritional value, etc., or which may also represent harmful residues.

The test methods necessary are generally complicated and need sophisticated equipment. However, there are also some simple and quick methods for chemical testing with sufficient accuracy which can be applied in the daily routine work such as pH-measurement, moisture/fat/protein determination and various screening methods utilizing test paper strips.

pH-measurement

The pH-value or acidity of meat is important in relation to the meat's microbiological and keeping quality. In the live animal the pH-value of the muscular tissue is about 7.0 to 7.1. Very soon after slaughter a drop in the pH-value is observed and after several hours (24 hours or less) the pH-value reaches its lowest level of about 5.6 to 5.8. The increasing acidity is because of the post-mortem formation of lactic acid from glycogen, a sugar-like substance stored in the live animal's muscles for energy supply.

In meat lactic acid causes a decrease in pH-value which is favourable for keeping quality (low pH inhibits bacterial growth) and for flavour (acidity is one of the components of meat flavour). However, the pH of meat is not uniform either in different carcasses or in different muscles of one carcass. Physiological oscillations do not greatly harm meat quality but abnormal reactions in meat are of great economic, hygienic and technological impact.

There are two types of abnormal reaction with regard to the pH in meat. First the pH-value may drop too fast and second it may not reach the normal low level several hours after slaughter, but remain in the range of 7.

Both abnormalities can easily be detected by pH-measurement in the meat. A too fast pH-value decrease is evident, when one hour after slaughter low pH-values in the range of 5.6 to 5.8 are already reached. This phenomenon occurs only in pigs and the meat remains pale, soft and exudative (PSE). Because of its paleness and wetness (low water-holding capacity), this meat should *not* be used for ham and sausage manufacture (gives dry, tasteless products).

An insufficient decrease of the pH-value, which occurs both in pork and beef, is of hygienic significance because of the lack of building up a certain degree of acidity and suppressing microbiological growth. This meat also remains close to pH-value 7 after several hours, and is dark, firm and dry (DFD). It should not be used for meat and meat products which have to be stored over a longer period, such as vacuum-packed meat cuts, dry sausages of the salami-type or cured raw hams. However, it is well suited for cooked meat products because of its extremely good water-holding capacity.

It can be seen from this that the pH-measurement is of particular importance for the selection of the raw material for meat processing purposes. Hence, portable electric pH-meters are widely distributed and utilized in the meat industry.

The pH is measured on meat surfaces or in the meat itself, in the latter case by pushing the sensor into the muscle or by means of an incision using a knife. The sensor consists of a glass electrode filled with an electrolyte (solution of KC1) and a sensitive glass membrane attached at the top.

Through the membrane the difference in the hydrogen-ion concentration, which corresponds to the acidity of the meat, is detected and digitally displayed on the attached instrument.

pH-measurement on meat can easily be performed but the following points must be considered:

- the electrode sensor must be completely filled with the electrolyte;
- the instrument must be adjusted daily (calibrated) using two buffer solutions with pH-values 4 and 7;

- after each measurement the electrode must be cleaned using distilled water;
- before each measurement the temperature of the meat, meat product, etc. must be checked and the instrument adjusted accordingly.

Moisture/fat/protein determination

Information about the moisture, fat and protein content is essential for the evaluation of the quality of different meats and meat products. Determination methods have changed a great deal in this field in recent years. Revolutionary techniques were introduced using X-rays, infra-red radiation or microwaves in automatic equipment for quick analyses of moisture, fat and protein. These modern methods are time-saving, the results are delivered within minutes or seconds and high numbers of samples can be tested. However, the equipment is expensive and therefore not suitable for small industries. For routine controls, where not necessarily highly accurate but reliable results on moisture, fat, protein and anorganic components (ash) are needed, cheaper and less complicated methods can be applied. A specially designed laboratory scale, together with some other devices, is required. After homogenizing and weighing the sample, it is fast dried using an infrared beam (or a microwave oven if available). The weight difference is equivalent to the product's water (moisture) content. The fat is then dissolved using a fat-extracting liquid and removed together with the liquid. The solvent is evaporated. The weight of the residue represents the fat content of the sample. Finally, the sample is charred in a muffle furnace and the weight of the residue is the ash content. Since the sum of the percentages of moisture, fat, ash and protein must be 100, and since the percentage of moisture, fat and ash is known, the protein content in percent is calculated as follows: 100% minus the percent of moisture, fat and ash. This method is not precise, but it is fast, provides useful results about the composition of meat and meat products and can be applied without high costs.

For chemical evaluations a number of screening methods are also available using different test papers. The results are indicated by changes of the colour of certain areas on the paper strips. These test papers are used for pH-measurement, screening of the nitrite content and even for the screening of some harmful residues such as antibiotics. pH-measurements on meat with test strips are negatively influenced by the meat pigment making the colour determination often difficult and the pH-determination not very accurate.

Microbiological Examination

These control methods cannot be carried out without laboratory equipment, because they require sample preparation under sterile conditions, incubation of the samples under constant temperatures and sufficient microbiological knowledge on the part of the personnel involved to interpret the results. However, the application of microbiological methods is the only way to obtain information about the hygienic status of places, equipment and foods. It is true that unclean conditions will always indicate high microbiological contamination and one could argue that a thorough cleaning-up rather than a further microbiological analysis would be needed in those cases. But there could also be the need of detecting the source of permanent contamination (for example through the water, movement of personnel, raw material delivered, etc.) or of food poisoning bacteria. Under these circumstances microbiological examinations can often be very helpful and solve immediate problems.

Some methods suitable for routine work should be mentioned.

Trigger methods

Microbiological culture media in special small moulds are lightly pressed against walls, equipment (knives, machines), meat surfaces or hands of personnel. The micro-organisms adherent to these objects are absorbed by the surface of the culture media, and after adequate incubation (one to two days at 30 to 37°C), microbial colonies can be identified and counted on the media. Each one of the colonies grown during incubation corresponds to one micro-organism which was on the object tested.

Instead of culture media a special sterile strip of cellotape together with a trigger can be used for taking samples from surfaces. After that the cellotape is laid on a culture media for incubation. This procedure allows the utilization of one culture medium for the incubation of different samples at the same time. However, there is one disadvantage with the trigger system. In the case of high bacterial contamination of the surfaces, tested bacterial colonies will grow very densely together and can no longer be counted.

Swab method

Surface contamination related to a certain area can be sampled using a sterile swab. After rubbing the swab gently along the surface to be tested, it is suspended on the surface of a culture media. In contrast with the trigger method, bacterial contamination can be spread over the whole surface which is important in the case of high contamination. Thus the samples can always be evaluated since the single colonies are not grown together. However, the method lacks some accuracy since bacteria may remain in the swab.

References

Bender, AK. and Zia, M. 1976. *Meat quality and protein quality*. 1. Fd. Technol. 11. 495498.

FAO. 1990. *Manual of simple methods of meat preservation*. FAO Animal Production and Health Paper No. 79. Rome, FAO.

Fennema, O. 1975. *Effect of freeze-preservation of food processing*. pp 244-288. Ed. R.S. Harris and Karmas, E. Avi Publishing Co. Westport, Conn.

Hudson, R.J. 1976. Potential for Meat Production from Marginal Land Resources. *Misc. Publ. Dept. of Animal Science*. University of Alberta, Edmonton, Alberta, Canada

Karmas, E. 1976. *Processed meat technology*. New Jersey, USA, Noyes Data Corporation.

4

Freezing and Refrigerated Storage in Fisheries

One of the important issues affecting fish preservation is the large biological variations existing from one region of the world to another and from one species of fish to another. This, combined with the fact that catching methods and consumption habits vary, has a considerable influence on the handling and preservation of the product.

In order to choose and operate refrigeration systems in the best possible way, some knowledge of the fish biology and factors influencing the quality are essential.

Composition of Fish

The fish has a skeletal or cartilaginous structure which provides support for the body. The muscles which form the edible part account for most of the weight of the fish. The skin forms a cover, often with an outer layer of scales, and secretes a slimy mucus, which lubricates the fish and seals the surface. The gills are the main part of the breathing mechanism and take up oxygen from the water. The organs in the body cavity, including the stomach, intestine and liver are known as the guts. Removal of the guts is normally the first step in handling and preservation. Shell fish has no backbone, but a hard outer cover or shell exoskeleton, which gives the necessary support and protection.

The principal components of the fish muscle - water, fat and protein - must be preserved with little or no changes. The protein content is usually in the region of 15-20 percent, whereas the fat content varies widely from species to species and from season to season. It can be as low at 0.5 percent in lean starved fatty fish and can reach over 20 percent in some species. In lean fish the bulk of the fat is stored in the liver and not in the muscle. Water is the main constituent, with considerable variations, typically 80 percent in lean fish and 70 percent in fatty fish. Carbohydrates, minerals, vitamins and some water extractable components are examples of other minor substances present.

Spoilage of Fish

As soon as a fish dies, spoilage begins. Spoilage of fresh fish is a rather complex process and is caused by a number of inter-related systems, some of which are suppressed by others. The factors which principally contribute to the spoilage are the degradation of protein with a subsequent formation of various products like hypoxanthine, trimethylamine, development of oxidative rancidity and the action of micro-organisms.

In live fish, food in the gut is reduced to simple substances, such as sugar and aminoacids, which are absorbed into the blood stream. The blood conveys these essential substances to sites where they are required, notably in the muscles. Production of these substances is induced by enzymes, which act as catalysts to chemical reactions, both in the gut and in the flesh. The enzymes remain active after death and thus bring out self-digestion, affecting the flavour, texture and appearance of the fish. After a fish dies, stiffening of the muscle called *rigor mortis* sets in and commences, due to the action of enzymes. Subsequently softening of the flesh occurs as self-digestion proceeds.

Self-digestion can take place rapidly in fish, especially in small fatty fish full of feed, where the gut enzymes are particularly active. The well-known phenomena "Burst Belly", which can occur in only a few hours after catch in sardines, herring and some other fish, is caused simply by a weakening of the belly wall due to self-digestion. The rate of selfdigestion is much dependent on temperature. Chilling of the fish to just above the freezing point does not stop, but retards self-digestion. Enzyme action can be stopped by heating; it is controlled to some extent by other methods, such as salting, frying, drying and marinating.

Micro-organisms are present in the surface slime, on the gills and in the intestines of the fish, but the muscle is sterile. Although it is not known with certainty how long it takes for the bacteria to penetrate the skin of the muscle three to four days is a reasonable estimate, but each species may be somewhat different. Fresh fish is seldom the cause of food poisoning, since the bacterial growth tends to make the muscles or flesh unpalatable before any toxins develop.

The environmental microflora introduced by cooling medium and by handling are also responsible for spoilage after the initial phase of self-digestion. Soon after the fish dies, bacteria will enter at a number of points, through the gills and into the blood vessels, through the lining of the belly cavity and eventually through the skin. Once in the flesh they can grow and multiply rapidly, producing disagreeable odours and flavours.

There are many different types of micro-organisms, each type having particular conditions for optimum growth. Thus it has been found that certain types of microorganisms dominate, depending on the initial infection, the properties of the food material, the temperature and other conditions. By cooling the fish to around 0°C, some of the bacteria groups responsible for the spoilage will cease to grow and the rate of spoilage will thereby be reduced.

Ambient conditions, such as the amount of moisture and oxygen available, have a marked effect on the microbiological activity. In melting ice the rate of fish spoilage is to some extent dependent on the rate of melting. Providing there is sufficient ice to maintain the desirable fish temperature of 0°C a higher melting rate can give slightly better results than a lower melting rate presumably due to the washing effect. Where the fish is in contact with surfaces such as wood, metal or other fish, foul odours can arise due to the action of certain anaerobic bacteria, which thrive in the absence of oxygen.

As microbiological action is the main and fastest cause of spoilage, great care must be taken to avoid conditions which accelerate the growth of micro-organisms. The growth rate is highly temperature-dependent and the principal preservative measure, besides good hygienic conditions, is to cool the fish as soon as possible after catching. Other supplementary measures have been tried, for example the use of antibiotics and different gases. So far only marginal improvements have been achieved and these supplementary methods have not found any wide applications.

Some of the chemical changes are caused by enzymatic reactions, the first taking place even before any serious changes are caused by microbiological activity. These enzymatic reactions are associated with *rigor mortis*. The result of those changes is that some constituents are chemically altered and some even disappear, altering the sensory properties -odour and flavour. Some of these substances, commonly known as extractives, are the first to be changed by the microbiological activity and the protein of the muscles will change considerably later.

The extractives are present in varying amounts from species to species. Herring and mackerel contain large amounts of amino-acid histamine, whereas cod and haddock only contain traces. Skate, dogfish and shark contain large quantities of urea which is absent in cod.

Trimethylamine oxide, which is available in all the salt water fish is usually absent from fresh water species. The breakdown of trimethylamine oxide into trimethylamine (TMA) is an important reaction, as the chemical determination of TMA may be used in quality assessment of salt water fish. Equally important is the determination of ammonia in some species, eg., sharks. Ammonia being formed during the breakdown of urea.

Chemical denaturation of proteins to a noticeable degree appears normally late in the deterioration process, as does oxidation of fat.

Development of oxidative rancidity is extremely variable in fresh fish. The ease with which some fish undergo oxidative rancidity is in part, explained by the large proportion of the highly unsaturated fats that many fishes contain. There is, however, a great difference between fatty species, such as mackerel and herring and fish such as cod and haddock. The former group have a high lipid content, free fat content and proportion of triglycerides, while the latter have a low lipid content, chiefly in the form of phospho lipids and lipoprotein immediately associated with muscle proteins. Even within a single fish itself there is a difference in the ease with which different portions undergo rancidity. Seasonal variations in susceptibility to rancidity have also been found.

Influence of Temperature

Fish begins to spoil immediately after death. This is reflected in gradual developments of undesirable flavours, softening of the flesh and eventually substantial losses of fluid containing protein and fat. By lowering the temperature of the dead fish, spoilage can be retarded and, if the temperature is kept low enough, spoilage can be almost stopped.

Rigor mortis, over a period of hours or days soon after death, can have a bearing on handling and processing. In some species the reaction can be strong, especially if the fish has not been chilled. The muscles under strain tend to contract, therefore, some of the tissue may break, especially if the fish is roughly handled, leaving the flesh broken and falling apart. If the muscles are cut before or during *rigor*, they will contract and in this way fillets from fish can shrink and acquire a somewhat rubbery texture. In many species, however, *rigor mortis* is not strong enough to be of much significance.

The freezing process alone is not a method of preservation. It is merely the means of preparing the fish for storage at a suitably low temperature. In order to produce a good product, freezing must be accomplished quickly. A freezer requires to be specially designed for this purpose and thus freezing is a separate process from low temperature storage.

What Happens During Freezing

Fish is largely water, normally 60-80 percent depending on the species, and the freezing process converts most of this water into ice.

During the first stage of cooling, the temperature falls fairly rapidly to just below 0°C, the freezing point of water. As more heat requires to be extracted during the second stage, in order to turn the bulk of the water to ice, the temperature changes by a few degrees and this stage is known as the period of "thermal arrest". When about 55% of the water is turned to ice, the temperature again begins to fall rapidly and during this third stage most of the remaining water freezes. A comparatively small amount of heat has to be removed during this third stage.

As the water in fish freezes out as pure crystals of ice, the remaining unfrozen water contains an ever increasing concentration of salts and other compounds which are naturally present in fish flesh. The effect of this ever increasing concentration is to depress the freezing point of the unfrozen water. The result is that, unlike pure water, the complete change to ice is not accomplished at a fixed temperature of 0°C, but proceeds over a range of temperature.

Literature on the freezing of fish is confusing and often contradictory about what happens to fish as it freezes. This is particularly the case when reference is made to the difference between slow and quick freezing. One of the main reasons for this apparent confusion is that only in recent years has knowledge of the freezing process advanced sufficiently to explain these differences in freezing rates. The result is that much of the literature still in circulation is now outdated.

At first it was thought that rapid freezing was unsatisfactory since sudden cooling could disrupt and tear the muscle tissue. It was also thought that, since water expands on freezing, it might be reasonable to expect the cell walls to burst under the induced pressure. There is some justification for both of these theories but they do not fully explain the differences between slow and quick freezing.

For some time a widely held view was that slow freezing resulted in the formation of large ice crystals which damaged the walls of the cells. This would then result in a considerable loss of fluid when the fish was thawed. The smaller ice crystals formed, when fish is frozen quickly, were thought to do little damage to the cell walls and, as a result, little fluid was lost on thawing. Difference in size of ice crystal probably accounts for some of the differences between slow and quick freezing, but is has been shown that this still does not provide a full explanation. The walls of fish muscle cells are sufficiently elastic to accommodate the larger ice crystals without excessive damage. Also, most of the water in fish muscle is bound to the protein in the form of a gel, and little fluid would be lost even if damage of the above nature did occur.

Slow freezing, however, does result in an inferior quality product and this is now thought to be due mainly to denaturation of the protein. Changes take place in some fractions of the protein as a result of freezing and since they are altered from their "native" state they may be said to be "denaturated", hence the term "protein denaturation". This denaturation depends on temperature and as temperature is reduced the rate of denaturation is reduced. Denaturation also depends on the concentration of enzymes and other compounds present. Thus, as the water is frozen out as pure ice crystals, the higher concentration of compounds in the unfrozen portion will result in an increase in the rate of denaturation. These two factors, which determine the rate of denaturation, act in opposition to each other as temperature is reduced and it has been demonstrated that the temperature of maximum activity is in the region of -1 to -2°C.

Slow freezing means that a longer time is spent in this zone of maximum activity and it is now thought that this factor accounts for the main difference in quality between slow and quick frozen fish.

Quick Freezing

There is no widely accepted definition of quick freezing.

It is unlikely that even a trained taste panel could detect the difference between fish frozen in 1h and 8h, but once freezing times begin to extend beyond 12h the difference may well become apparent. Freezing times of up to 24h or even longer, achieved in some badly designed and operated freezers, will almost certainly result in an inferior product. Very long freezing times, for example, due to freezing fish by bulk stacking in a cold store, may even result in spoilage by bacterial action before the middle of the stack is sufficiently reduced in temperature.

Since the temperature just below 0°C is the critical zone for spoilage by protein denaturation, an early UK definition of quick freezing recommended that all the fish should be reduced from a temperature of 0°C to -5°C in 2h or less. The fish should then be further reduced in temperature so that its average temperature at the end of the freezing process is equivalent to the recommended storage temperature of -30°C. With normal freezing practice in the UK, this latter requirement is defined by stating that the warmest part of the fish is reduced to -20°C at the completion of freezing. When this temperature is reached, the coldest parts of the fish will be at, or near, the refrigerant temperature of say -35°C and the average temperature will then be near -30°C. This is a rather elaborate definition of quick freezing and it is probably more strict than is necessary to ensure a good quality product.

The more widely used definitions of quick freezing do not specify a freezing time or even a freezing rate but merely state that the fish should be frozen quickly and reduced in the freezer to the intended storage temperature.

Double Freezing

Double freezing means freezing a product, thawing or partly thawing it, and refreezing. This practice is often necessary for the production of some frozen fish products made from fish previously frozen and stored in bulk. What must be remembered is that even quick freezing results in quality changes in the fish and double freezing will therefore result in further changes. Only fish that were initially very fresh could therefore be subjected to double freezing and still conform to good quality standards. Fish frozen quickly at sea immediately after catching, for instance, would be suitable for this purpose.

Handling of Fish before Freezing

Freezing and cold storage is an efficient method of fish preservation but it must be emphasised that it does not improve product quality. The final quality depends on the quality of the fish at the time of freezing as well as other factors during freezing, cold storage and distribution. The important requirement is that the fish should at all times be kept in a cool condition before freezing, about 0°C, and the use of ice or other methods of chilling is recommended.

Frozen Fish

Freezing and frozen storage of fish can give a storage life of more than one year, if properly carried out. It has enabled fishing vessels to rem ain at sea for long periods, and allowed the stockpiling of fish during periods of good fishing and high catching rates, as well as widened the market for fish products of high quality.

The mechanism by which frozen fish deteriorates is somewhat different from that causing spoilage of chilled fish. Provided the temperature is low enough - below -10°C bacterial action will be stopped by the freezing process. Chemical, biochemical and physical processes leading to irreversible changes will still occur, but at a very slow rate. Deterioration during frozen storage is inevitable, and in order to obtain satisfactory results, fish for freezing must be of good quality.

The proteins changes in fish frozen under poor conditions can be recognised in the thawed fish. The normally bright, firm and elastic product becomes dull and spongy. The flesh will tend to sag and break and there will be substantial losses of fluid, which can be squeezed out easily. When cooked the fish will be dry and fibrous. The rate at which protein denaturation takes place in frozen fish depends largely on the temperature and will slow down as the temperature is reduced.

Changes taking place in the lipids of the frozen fish will also slow down when the temperature is reduced. The oxidation of the fat leads to objectionable flavours and odours. This can be particularly serious in fish of high fat content and probably also accounts for most of the flavour changes in lean fish. Some substances, notably salt, and some processes, such as drying, can aggravate the problem. Smoked fish, for example, has a shorter storage life in frozen condition than the raw, frozen counterpart. The addition of chemicals to prevent oxidation has not been successful, except for some special types of products.

The rate of oxidation can be reduced by reducing the exposure to oxygen. This can be achieved by introducing a barrier at the surface of the fish. Thus fish in a block keep better than fish frozen individually, and the addition of an ice glaze is beneficial. Glazing is carried out after freezing by brushing or spraying chilled water onto the surface of the fish or by dipping in cold water. Packaging materials, impermeable to moisture and oxygen can be effective, especially if vacuum packaging is employed.

Some transfer of moisture from the product is unavoidable during freezing and frozen storage, which leads to dehydration of the fish. Good operating conditions are essential in order to keep dehydration to a minimum. It has been clearly established that fluctuating cold store temperatures are a major cause of dehydration. In practice the more severe cases of drying occur during frozen storage rather than during freezing. In extreme dehydration the frozen fish acquires a dry wrinkled look, tends to become pale or white in colour and the flesh become spongy. This characteristic appearance is called, inappropriately, 'freezerburn'. The weight loss is, of course, serious from an economic point of view and dehydration will accelerate the other important changes - protein denaturation, as well as oxidation. Glaze on the exposed surfaces of the fish before storage will however, evaporate over a period of time and drying of the fish itself will resume. Reglazing is therefore a common need. Paper wrappers can be used as a protection, but depending on the conditions some drying of the fish within the packing will still occur.

Frozen Fish Products

The variety of species, processes, methods of presentation and packaging available provide scope for the preparation of numerous frozen fish products. These products, however, can be separated into two main groups; products intended for direct consumption and products intended for further processing.

Products for Direct Consumption

Individually quick frozen (IQF) products are frozen as single units which need not be thawed for sub-division or perhaps even for cooking purposes. IQF single fillets and shrimp are two products of this type.

The demand for IQF products has increased with the upsurge in the number of low temperature "freezer" cabinets both in catering establishments and in the home. IQF freezing allows for the purchase of a frozen product in bulk and the selection from storage of only sufficient quantities to meet immediate requirements.

Other products such as blocks of fish and fish portions usually packaged in cartons are also produced for direct consumption without the need for reprocessing. The consumer will purchase this type of product from the retailer, still in the frozen state, and either cook it in the frozen state or thaw it for immediate consumption.

The production of products for direct consumption may not yet be appropriate in many developing countries. This type of product requires the provision of an extensive network of refrigerated storage and transport. This facility, which is popularly known as the "cold chain", may not be developed enough to enable this system to operate.

Products for further processing

These products are produced for two purposes:

1. Frozen in bulk and thawed after storing, to be used as newly caught, unfrozen fish.
2. Frozen in bulk and after storage, further processed without thawing so that it may be presented as a retail pack.

Products frozen in bulk can be unprocessed, such as blocks of whole fish frozen in contact freezers. Blocks of frozen fish may weigh up to 50 kg; they are usually glazed or wrapped after freezing and are then stored until required for further processing.

In some cases, fish are bulk frozen, stored and finally thawed all in one place. This is usual when there is a short seasonal fishery and fish are preserved for processing over a longer period. Bulk frozen fish may also be distributed in the frozen state. This enables the fish to be sold to a larger home market and also allows the product to be exported. In this case there are additional requirements for low temperature transport and a more extensive cold chain.

Fish frozen in bulk may also be fully processed before freezing and only the skinless, boneless portion used. One particular process of this type worth special mentioning is the production of frozen fillet blocks. A frozen fillet block is a regular shaped block of fish flesh frozen in a horizontal plate freezer within a treated cardboard carton and a metal retaining frame. The filling process ensures that there are no voids in the block. After freezing, the blocks are stored in bulk and at a later date cut into smaller portions of different shapes. The fish portions may then be packaged and sold in this form or they may be coated with a flour batter and breadcrumbs. Coated fish portions should be returned to the freezer and rehardened before packaging and storing.

The type of frozen fish product and the form in which it is produced in a particular country may well depend on the extent of the cold chain as well as on the demands of the consumer. It therefore seems likely that in most developing countries a bulk freezing process will be the initial development. This will enable the industry to cater for seasonal variations and allow a wider distribution of the fish catch. Other frozen products will follow later when the industry develops and the cold chain is extended.

FREEZERS

There are now many different types of freezer available for freezing fish, and freezer operators are often uncertain about which type is best suited to their needs. Three factors may be initially considered when selecting a freezer; financial, functional and feasibility.

Types of Freezers

The three basic methods of freezing fish are:

1. Blowing a continuous stream of cold air over the fish - air blast freezers.
2. Direct contact between the fish and a refrigerated surface - contact or plate freezers.
3. Immersion in or spraying with a refrigerated liquid - immersion or spray freezers.

Air Blast Freezers

The advantage of the blast freezer is its versatility. It can cope with a variety of irregularly shaped products and whenever there is a wide range of shapes and sizes to be frozen, the blast freezer is the best choice. However, because of this versatility it is often difficult for the buyer to specify precisely what he expects it to achieve and, once it is installed, it is all too easy to use it incorrectly and inefficiently.

The use of air to transfer heat from the product being frozen to the refrigeration system is probably the most common method used in commercial refrigeration. The natural convection of the air alone would not give a good heat transfer rate, therefore, forced convection by means of fans has to be introduced. To enable the product to be frozen in a reasonable time the air flow rate should be fairly high. Also, in order to obtain uniform freezing rates throughout the freezer, the air flow requires to be consistent over each fish or package.

Nearly all air blast freezers operate with finned tube coolers. The fins greatly extend the surface for heat exchange, and the closer the fins the greater will be the surface area and the smaller the cooler unit. Moisture lost from fish during freezing and from air infiltrating into the cooler will eventually be deposited as frost on the cooler surface. If this frost eventually bridges the space between the fins, the effective cooler surface is then reduced, the rate of heat transfer will be reduced and the freezer temperature will rise. There will also be a greater resistance to air flow through the cooler and the air flow rate may be reduced.

Most of the water lost from the fish is lost during the early stages of freezing and in some freezer designs, this will mean a higher degree of frosting on some parts of the cooler than on others. This will effectively reduce the period of operation before a defrost is necessary. Frost build-up on the cooler is also more prolific on the front, upstream coils; therefore a cooler with a large frontal area will be able to operate longer before a defrost is necessary. The specified fin spacing may also be increased where there is likely to be a quick build-up of frost. A good freezer design should be able to operate for at least 8h before a defrost is required but a poor design may require defrosting every 2h.

There are many different designs of air blast freezer both for batch and continuous operation. Details are given of a number of types of air blast freezer in common use, with comment on their suitability for various products and methods of processing and also on their limitations.

Continuous air blast freezers

In this type of air blast freezer, the fish are conveyed through the freezer (on trucks or trolleys or they may be loaded on a continuously moving belt or conveyor) usually entering at one end and leaving at the other.

When trucks or trolleys are used, they are loaded at one end of the freezer and progressively moved along the freezer as additional trucks are loaded. Once the freezer is full, a truck has to be removed from the exit end before a fresh truck can be loaded. This batch-continuous operation must always allow the coldest air to flow over the coldest fish; otherwise fish which are well frozen will be subject to warmer air as new trucks are loaded. One difficulty with this type of freezer is that when the freezer is fully loaded, a whole row of trucks has to be moved at one time. This is particularly difficult at very low temperatures since special bearings and lubricants are required for the truck wheels and it is difficult to keep the trucks free of frost and ice. Trolleys have been suspended from overhead rails to overcome some of these difficulties but this equipment is cumbersome and still not easy to operate.

To avoid moving trucks within the freezer, a batch-continuous freezer can be designed with a cross flow air arrangement and the freezers may then be loaded from the side. Again in this freezer, once it has been fully loaded, a truck is removed before a fresh one is added. It is a simple matter to keep account of the loading sequence of the freezers by having hand-set clock dials above each entrance which will indicate the time the truck or trolley will be ready for unloading. This cross-flow arrangement allows a cooler with a large frontal area to be built, and frost is also deposited uniformly.

Continuous air blast freezers using belts or conveyors for moving the product through the freezer can only be used if the product can be frozen quickly. It is unlikely that a product with a freezing time of more than 30 min would be suitable for this freezer. The reason for the limitation on freezing time is that the freezer will become too long and cumbersome if a long freezing time is required. The freezing time, the freezing requirement in kg/h and the loading density of the product on the belt determine the freezer dimensions.

Continuous belt freezers generally have their own special problems. The belt has to be flexible, easily cleaned, non corroding, suitable for use in direct contact with food and should not interfere unduly with either the freezing time or adversely affect product quality. Stainless steel mesh link belts or chain link belts are mainly used for this purpose but they have certain disadvantages. Apart from being expensive, they affect the appearance of the product. If fish are loaded directly on the belt, the crinkled or indented appearance of the frozen product is not always acceptable. Open mesh belts can also give rise to difficulty when removing the product after freezing, and some weight loss may be incurred due to slight physical damage. Skin-on fillets call usually be removed quite easily but skinless fillets and fish portions can stick to the belt and cause unacceptable weight losses.

Plastic belts made in the form of interlocking links have been used in some continuous freezers. These belts, add about 10 percent to the freezing time. They suffer from the same indentation problems as metal mesh belts but transfer is generally easier. However, their larger mesh makes them unsuitable for small products. If they are only used for the initial part of the freezer, the fish can be surface-hardened and then be transferred to a stainless steel belt. This would allow a two-belt operation in the freezer. In spite of these often minor difficulties in obtaining an ideal belt for continuous belt freezers, many are successfully operated for freezing a variety of products.

Continuous belt freezers can be constructed with either cross-flow or series-flow air circulation. In the series-flow arrangement, the direction of air flow must be such that the coldest fish meet the coldest air. The design of the belt entry and exit must keep the rate of air infiltration to a minimum.

In a continuous freezer, there is no scope for rearranging the volume or space for different products. The belt speed, however, is usually variable and this can be adjusted to accommodate different product freezing times. The capacity of a continuous freezer can therefore vary considerable depending on the product being frozen and its freezing time

Another important consideration when using a continuous air blast freezer is whether the freezer will be used continuously. A continuous freezer left in operation but not fully loaded could give rise to higher freezing costs per kg of product frozen.

Batch air blast freezers

Batch air blast freezers use pallets, trolleys or shelf arrangements for loading the product. The freezer is fully loaded, and when freezing is complete, the freezer is emptied and reloaded for a further batch freeze. Apart from this difference in mode of operation, the batch freezer gives rise to bigger fluctuations in the refrigeration load than continuous or batch-continuous freezers.

This large fluctuation in refrigeration load means that the refrigeration system will require special control arrangements to cater for the variations. Capacity control or a multiunit system can be used or a competent engineer can manually control the system to match the load. Some refrigeration systems are also better suited to this type of variable load application than others.

It is seldom that fish processing can be arranged so that all the fish can be loaded into a batch freezer at the same time. Therefore, if each trolley or pallet is loaded as and when it is ready, the refrigeration peak load will be considerably reduced. This will make the operation similar to a batch-continuous process, but again, care should be taken not to place warm fish upstream of a partly frozen product.

Fluidized and semi-fluidized freezers

One type of air blast freezer fluidizes the product with a strong blast of air from below. The product then behaves like a fluid and when poured into the trough at the input, it moves along the length of the freezer without mechanical assistance and over-flows at the output. This type of freezer has been used successfully for such products as garden peas which are readily separated and kept apart but, as yet, the freezer has not had a wide application for fish or fishery products. Small cooked and shelled shrimp is one of the few fish products that has been successfully frozen by this method.

A modified fluidized freezer which may be termed a semi-fluidized freezer has also been used for fish-freezing applications. A conventional conveyor is used but at the early stages of freezing, sufficient air is blown from below the belt to agitate the product and ensure that individual portions remain separate until the outer surface has been hardened. This type of freezer can be used with a double belt, with transfer from one to the other midway through the freezing process.

There is however some difficulty in judging the correct air flow to produce the slight agitation required and a fixed flow rate is not suitable if a variety of products are to be frozen. Also, with many products there still remains some difficulty in making the transfer from one belt to the other.

Loading a batch air blast freezer

Because of their versatility, batch air blast freezers are often misused by operators who do not realise their freezing limitations.

The size of the refrigeration plant is fixed to match a given freezing requirement at the designed freezer operating condition. However, if the freezer is used for freezing other products which have different space requirements and freezing times, the freezer operating condition will change. Depending on the original design specification, the freezer may therefore be overloaded or underloaded by a change in product.

Good performance in batch air blast freezers is obtained by freezing the product in open trays without wrapping. Trays used in air blast freezers should transfer heat readily, be easily emptied and also be robust. Normally they are required to produce a pack that is of regular shape but when the product allows their use, trays with a taper on the sides of about one in eight can be emptied by applying a cold water spray on the underside for a few seconds and then giving a gentle tap on the edge. Trays used in this manner should never be filled above the tray edge or the product will be damaged during release.

Cleaning and drying of trays before re-use is necessary to maintain a high standard of hygiene. Where the rate of production justifies the cost, an automatic tray washer may be installed.

Plate Freezers

Plate freezers and air blast freezers are the types of freezer most commonly used for freezing fish in industrial countries. Plate freezers do not have the versatility of air blast freezers and can only be used to freeze regularly shaped blocks and packages.

Plate freezers can be arranged with the plates horizontal to form a series of shelves and, as the arrangement suggests, they are called horizontal plate freezers (HPF). When the plates are arranged in a vertical plane they form a series of bins and in this form they are called vertical plate freezers (VPF).

Modern plate freezers have their plates constructed from extruded sections of aluminium alloy arranged in such a manner as to allow the refrigerant to flow through the plate and thus provide heat transfer surfaces on both sides. Plate freezers are fitted with hydraulic systems which move the plates together and apart.

Horizontal plate freezers

The two main uses for this type of freezer are the freezing of prepacked cartons of fish and fish products for retail sale and the formation of homogeneous rectangular blocks of fish fillets, called laminated blocks, for the preparation of fish portions. The thickness of package or block frozen is 32 to 100 mm and the freezer can readily adapt from the thicker to the thinner package provided the range required is made known to the supplier at the time of purchase. There is no direct contact between the fish and the freezer plates when freezing by this method since the fish is always packaged before freezing. If the operator is also careful not to spill water on the plates during loading and unloading, the freezer may be operated with only a light brush between each freeze to remove surface frost. The door may be left open overnight to allow the plates to defrost fully after being hosed down with warm water. A hot gas defrost arrangement is the quickest method to defrost an HPF, but even with this method, it may take 30 min or more. The defrosted plates must be completely free from frost or ice and dried before the freezer is used again.

Horizontal plate freezers intended to be operated with a hot gas defrost are fitted with additional pipework which allow the cold refrigerant to be discharged from the bottom of the freezer as the defrost proceeds. Without this special pipework and operating valves, a hot defrost would clear the top plates only and leave the cold refrigerant in the plates at the lower levels. As in all hot gas defrost systems, the refrigeration system must have an adequate load to provide sufficient hot gas for an effective defrost. This system would therefore be better applied when there are two or more freezers operated from a common refrigeration system and each freezer will then be defrosted in turn while the others are in operation.

Vertical plate freezers

The main advantage of this type of freezer is that fish can be frozen in bulk without the requirement to package or arrange on trays. The plates form what is in effect a bin with an open top and fish are loaded directly into this space. This type of freezer is therefore particularly suitable for bulk freezing and it has also been extensively used for freezing whole fish at sea. The maximum size of block made by this method is usually 1 070 mm x 535 mm. Other dimensions however, can be produced in which the thickness can vary from 25 to 130 mm, but will depend on the fish to be frozen. The maximum weight and dimensions are also limited by the physical effort required from the operator to lift the block, and by the ease with which it can be handled so that damage to the fish is kept to a minimum.

In most cases, fish can be loaded between the plates without wrappers and water need not be added either to strengthen the frozen block or improve the contact with the plates. Fish such as cod and haddock produce compact blocks with a block density of approximately 800 kg/m^3.

With fatty fish such as herring, it has been found advantageous to use wrappers and add some water to fill the voids in the block. Fatty fish do not form blocks which are as firm and strong as blocks made from lean fish especially during seasons when the oil content of the fish is high. Water added helps to strengthen the block, protects the fish during subsequent handling and reduces the effects of dehydration and oxidation during cold storage. Well formed, rigid blocks are particularly important when freezing at sea. The product may be handled under particularly adverse operating conditions and poorly formed blocks, prone to breakage, would result in a high percentage of loose fish. Machine filleting or splitting of the fish for instance, may be difficult if fins and tails are broken. Wrappers have been used when freezing fatty fish in VPFs to protect the exposed fish on the outside of the block. A wrapper that has been found suitable for this purpose is a single layer paper bag, coated internally with polyethylene, and shaped to fit the space between the freezer plates. Wrappers made from polyethylene with a specially roughened outer surface to reduce slippage have also been used.

Fish frozen in wrappers require a longer freezing time due to the insulating properties of the wrapping material. Some types of wrapper would have a considerable effect on freezing time but in sea trials the material described did not increase the freezing time by a significant amount.

Vertical plate freezers are defrosted to release the blocks of fish after each freeze. Fish are in direct contact with the plates and the force required to release the blocks without a defrost could be excessive and result in plate damage. The defrost time need not exceed 3 or 4 min if a suitable supply of defrost gas or hot liquid is available. If a primary refrigerant is used in the plates, a hot gas defrost is generally used. Where there is a multiple installation, the freezers are defrosted in turn with the other units in operation providing the necessary refrigeration load for the compressor. When a secondary refrigerant is used, a reservoir of hot liquid has to be maintained and pumped through the plates to displace the cold liquid present. With this arrangement, it is possible to return the bulk of the cold liquid to the low temperature reservoir at the start of defrost, and also return the warm defrost liquid to the hot liquid reservoir for reheating at the start of the next freeze. This arrangement reduces the quantity of liquid interchanged at each defrost but provision must be made to maintain the liquid charges in both the cold and hot systems at the correct level.

Defrost arrangements such as those described lead to more complicated and expensive refrigeration pipework. Attempts have been made to assist the release of the blocks by coating the plates with a low friction plastic material so that a defrost was unnecessary. Although this worked reasonably well, a defrost was found to be essential to prevent fish sticking to the plates which are at a temperature below 0°C, and thus failing to form a compact block. Freezing times are longer due to the poor contact being made with the plates and because of the lower block density, more storage space is required for a given quantity of fish.

Automatic plate freezers

This type of freezer freezes fish in cartons and is a continuous form of the HPF. Automatic plate freezers are specially designed for a processing line; and units with capacities of up to 2 t/h are available. Their main advantage is that they save the labour required for the loading and unloading of batch plate freezers. However, when this labour saving is related to the total labour requirement for packing and other operations, the saving is often not significant.

Liquid Nitrogen Freezer

In this freezer, the product is brought into direct contact with the refrigerant.

The fish on the stainless steel conveyor belt initially come into contact with the counter current flow of nitrogen gas at a temperature of about -50°C. As the fish. progress through the precooling stage of the freezer, the gaseous nitrogen partially freezes the fish and up to 50 percent of the product heat is extracted. The product then passes below the liquid spray where freezing is completed by the boiling liquid. The last stage in the freezer provides a few minutes for the fish temperature to reach equilibrium before the fish are discharged.

The main advantage of the liquid nitrogen freezer LNF is that freezing is very quick and the physical size of the freezer is correspondingly small. The freezer is operated without the need for compressors, condensers or coolers; therefore maintenance requirements are minimal and the power required to operate the freezer is very low. Liquid nitrogen must be retained in a vacuum insulated pressure vessel with continuous venting to keep the contents cool and the internal pressure down. One estimate given is that 0.5 percent of the stored contents is lost each day by this method. In addition, about 10 percent has been estimated to be lost during the transfer of liquid from the tanker to the storage vessel although the customer is not charged directly for this loss. This method of freezing is more expensive than most others, being up to four times more costly than conventional air blast freezing. Although the freezer is small and there is no refrigeration machinery requirement, storage space and access is required for the liquid nitrogen tank. The main disadvantage of this type of freezer in most developing countries is that delivery of nitrogen could be expensive and there may be no guarantee of regular supplies.

Carbon dioxide freezer

This type of freezer has been known for a long time and uses liquefied carbon dioxide which is usually a by-product of another industrial process.

The liquefied carbon dioxide is injected into the freezer and comes into direct contact with the product. In this respect, it is similar in operation to an LNF. With large units, it is economically feasible to recover the carbon dioxide and about 80 percent of the refrigerant used can be reliquefied. Carbon dioxide can be contained in insulated vessels at a moderate pressure and losses during storage are therefore negligible. High levels of carbon dioxide in the factory air are dangerous, therefore a freezer using this refrigerant must be vented and the gas discharged outside the building. Again, as is the case with other types of freezer which rely on regular supplies of refrigerant, carbon dioxide freezers would not be suitable for use in remote areas.

Immersion freezers

By using a liquid for the removal of heat from a product, favourable freezing rates can be achieved. Liquid can remove more heat per unit volume than gas (eg. air) but, like gas, a stagnant boundary layer is formed which slows the transfer of the heat. Liquids used for heat transfer must therefore be circulated over the product. Difficulties due to high viscosity often arise when a low temperature liquid is used.

Many liquids that have suitable refrigeration and heat transfer properties are not allowed to be used in direct contact with food. Those that are available are limited in their use because they may cause changes in texture and taste in the food with which they are in direct contact. Immersion in sodium chloride brine was one of the very first methods used to freeze fish since it was a logical progression from the method used to freeze block ice. Brine immersion freezing may still be used for such fish as tuna which are intended to be marketed as a canned product. The fish are large and have a thick skin; therefore the uptake of salt is not great. The little salt that is absorbed is not detrimental to the canned product since salt is usually added to the product before canning in any case. For many other fish freezing applications, adverse effects on texture and taste of the fish due to the absorption of brine have proved to be unacceptable. Even without excessive brine uptake, the surface of the fish will be coated and handling the product after freezing is difficult and messy. Some fish products such as shrimp have been frozen in syrup and salt solutions, and sugar and salt solutions but again there is some degree of absorption with changes in flavour.

Freezing Operating Temperatures

Bearing in mind that the freezer must reduce the temperature of the product to the intended temperatures of storage, freezers should operate at temperatures which allow this to be accomplished under the most favourable economic conditions. When selecting the appropriate freezer operating temperatures, account should also be taken of cost of equipment, operating costs, space requirements, quality considerations and other factors. In some types of freezer, the temperature is fixed by the method of operation, whereas in others, such as air blast and plate freezers, there is scope for varying the temperature to suit any particular requirement.

The following table gives some typical operating temperatures for various freezers:

Table 7. Freezer operating temperature

Type of freezer	*Operating temperature (°C)*
Batch air blast	-35 to -37 air
Continuous air blast	-35 to -40 air
Batch plate	-40 refrigerant
Continuous plate	-40 refrigerant
Liquid nitrogen	-50 to -196 refrigerant
Liquid carbon dioxide	-50 to -70
Sodium chloride brine	-21 refrigerant

Space Requirements for Freezing

The space required for a freezer obviously depends on the capacity and type of freezer. Some factors affecting total freezer space required are given below.

It can generally be assumed that, for a given capacity requirement, the quicker a freezer can freeze the product the smaller will be the physical space required. Freezer space, including that required for loading and unloading the product, is only one factor to be taken into account when calculating the total area requirement. Distinction should be made between floor space required within a building and that required in an open yard outside the covered factory area. Space is required for refrigeration machinery and access for maintenance but for small units, the machinery may be located above or below the freezer unit and will not add to the floor area. With liquid nitrogen and carbon dioxide freezers, no mechanical refrigeration is required, but storage must be made available for the refrigerant. In addition, an area has to be made available for manoeuvring the tanker supplying the refrigerant.

A working area is also required for handling and possibly packaging the product before and after freezing. Trolleys and pallets also require space and if they are doubled up to allow for a rotation system to be used, the floor area occupied by this equipment can be considerable. Packaged products also require a dry area for storing the packaging material which is often printed or marked to identify the product and the company, and this often means ordering in larger quantities.

Freezing Time

The freezing time is the time taken to lower the temperature of the product from its initial temperature to a given temperature at its thermal centre. Most freezing codes of practice require that the average or equilibrium temperature of the fish be reduced in the freezer to the intended storage temperature. The final temperature at the thermal centre is therefore selected to ensure that the average fish temperature has been reduced to this storage value. The recommended storage temperature for frozen fish in the UK for a period of 1 year is -30°C and, to ensure that the fish are frozen quickly, the temperature of the freezer must be lower than this.

The surface of the fish in a freezer will be quickly reduced to near the freezer temperature of say -36°C. Thus when the warmest part at the thermal centre is reduced to 20°C, the average temperature of the fish will be close to the required storage temperature of -30°C. The freezing time, in this particular case, will therefore be defined as the time taken for the warmest part of the fish, at the thermal centre, to be reduced to -20°C.

Variables which Affect Freezing Time

The factors which determine the overall heat transfer coefficient and hence the freezing time are listed below:

— *Freezer type*. The type of freezer will greatly influence the freezing time. For example, due to improved surface heat transfer, a product will normally freeze faster in an immersion freezer than in an air blast freezer operating at the same temperature.

— *Operating temperature*. The colder the freezer, the faster the fish will freeze. However, the cost of freezing increases as the freezer temperature is reduced, and in practice, most freezers are designed to operate only a few degrees below the required storage temperature of the product. For example, plate freezers usually operate at about -40°C and blast freezers at about -35°C when the storage temperature is -30°C.

— *Air speed in blast freezers*.Freezing time is reduced as the air speed is increased. This, however, is a rather complicated relationship and it depends on a number of factors. If the resistance to heat transfer of the stagnant boundary layer of air is important, changes in air speed will make a significant difference to the freezing time. If, however, the package is large and the resistance of the fish itself is the important factor then changes in air speed will be less significant. Air temperature, air density, air humidity and air turbulence are other factors that have to be taken into account when the effect of air condition on freezing time is considered. Some of these factors however, may only have a minor effect.

— *Product temperature before freezing*. The warmer the product, the longer it will take to freeze. Fish should therefore be kept chilled before freezing both to maintain quality and reduce freezing time and refrigeration requirement. For example, a single tuna 150 mm in diameter frozen in an air blast freezer will take 7h to freeze when the initial temperature is 35°C but, only 5h when the temperature is 5°C.

— *Product thickness*. The thicker the product, the longer is the freezing time. For products less than 50 mm thick, doubling the thickness may more than double the freezing time whereas doubling a thickness of 100 mm or more may increase the freezing time fourfold. The rate of change of freezing time with thickness therefore, depends on the relative importance of the resistance of the fish to heat transfer.

— *Product shape.* The shape of a fish or package can have a considerable effect on its freezing time and this is dependent on the ratio of surface area to volume.

— *Product contact area and density.* In a plate freezer, poor contact between product and plate results in increased freezing time. Poor contact may be due to ice on the plates, packs of unequal thickness, partially filled packs or voids at the surface of the block. Surface voids are often accompanied by internal voids and this also results in poor heat transfer. Apart from increasing freezing time, internal voids also reduce the density of the block.

— *Product packaging.* The method of wrapping and the type and thickness of the wrapping material can greatly influence the freezing time of a product. Air trapped between wrapper and product has often a greater influence on the freezing time than the resistance of the wrapping material itself.

— *Species of fish.* The higher the oil content of the fish the lower is the water content. Most of the heat extracted during freezing is to change the water to ice; therefore, if there is less water, then less heat will require to be extracted to freeze the fish. Since the fat content of oily fish is subject to seasonal variations, it is safer to assume the same heat content figure used for lean fish in any calculation. This also ensures that the freezer capacity is adequate whatever the species of fish being frozen.

Treatment of Fish After Freezing

As soon as fish are removed from a freezer, they should be glazed or wrapped and immediately transferred to a low temperature store. When it is known that storage will be for a short period only, glazing or wrapping may not be necessary or practical. Blocks of whole cod, frozen at sea, are usually transferred to the ship's cold store without a protective wrapper or glaze but this may be added later, prior to long term storage on shore. During relatively short terms of storage, fish without a protective wrapper or glaze can be severely dehydrated in a poorly designed or operated store.

Glazing

The application of a layer of ice to the surface of a frozen product by spraying, brushing on water or by dipping, is widely used to protect the product from the effects of dehydration and oxidation during cold storage. The ice layer sublimes rather than the fish below and it also excludes air from the surface of the fish and thereby reduces the rate of oxidation. Heat added by the glazing process is often considerable and the fish may require to be recooled in a freezer before being transferred to the cold store.

In order to form a complete and uniform glaze on the surface of the fish, the glazing process requires to be closely controlled. The amount of glaze applied depends on the following factors:

— Glazing time
— Fish temperature
— Water temperature
— Product size
— Product shape

Glazing by dipping in a container of water is not recommended. The initial temperature of the water may be relatively high; this is reduced as glazing proceeds and the thickness of glaze will therefore vary. The glaze on IQF fillets has been shown to vary between 2 and 20 percent using this method, even when the immersion time was kept constant. In practice, the time will not be constant and this will give rise to even greater inconsistency. The water will also become contaminated after some time; therefore, this method is not recommended. If a dipping method is used to apply a glaze, the container should be continuously supplied with chilled water and fitted with an overflow.

Spray glazing methods are suitable, but again it is difficult to obtain a completely uniform glaze and it may be necessary to invert the fish to ensure that all surfaces are treated.

The glazer has a number of features which enable a complete uniform glaze to be applied.

1. A constant speed belt will ensure a fixed time in the glazing zone;
2. The overhead and underside spray provide a constant supply of chilled make-up water and glazes both the upper and lower surface of the product;
3. The double belt arrangement forces fish to invert providing an even glaze.

4. The adjustable baffle may be used to rearrange overlapping fish on the belt so that each fish is totally exposed;

Glazing when the fish surface temperature is at -70°C or lower, e.g. immediately after cryogenic freezing, results in a glaze which is fractured and broken due to thermal stress during the formation of the ice. This glaze is easily dislodged during subsequent handling. If the fish are immersed in the glaze water for too long, a thick glaze is formed but the equilibrium temperature of fish and ice is high and only slightly below 0°C. This glaze will be soft and easily dislodged during subsequent handling.

Packaging of Frozen Fish

After fish has been frozen, it can be subjected to many forms of deterioration between production and eventual consumption. Contamination from humans, animals, insect and atmospheric sources is possible. Physical damage can be caused by bad handling during stacking, transit and storage or display in freezer cabinets. The sensory properties can be adversely affected by tainting and textural and flavour changes can be caused by dehydration and chemical changes which can take place under poor cold storage conditions. To prevent or reduce losses in product quality, it is essential that the frozen product is packaged in such a way as to provide an effective barrier with sufficient impact and compression strength to prevent mechanical damage. The packaging material must have adequate barrier properties to reduce losses due to dehydration and pick-up of taints. Further considerations are the printability of the material, so that a well designed attractive illustrated package can be produced. The package should give the consumer information on the nutritional properties of the product and instructions on how the product should be prepared, stored and give the 'use by' or 'best before' date.

A final consideration which is becoming increasingly important is the environmental issue. Considerations should include the impact of the packaging material on the environment, whether or not the package is re-usable or recyclable, whether the package is made from renewable resources and if the package produces pollution when it is being destroyed.

Types of Packaging Material for Frozen Fish

The range of packaging material for frozen fish is very wide and is dependant on the form of the product being packed. Whole fish frozen in a vertical plate freezer, for example, may require little packaging, other than an ice glaze, to prevent dehydration. Small pelagic fish frozen in vertical plate freezers may benefit from being frozen in water blocks where the freezer plate is lined with a robust plastic paper lined bag. This can be filled with water after the fish have been placed between the freezer plates. A processed fish product, for example fish sticks, may be wrapped in a primary package which is in direct contact with the frozen food and then stored in an outer carton. A number of primary packages may be collected together and packaged in a master carton (secondary package) for delivery or display in a freezer cabinet. The secondary packages can be bought together in a tertiary package, for example a wrapped pallet, and used for bulk transportation.

Primary Packaging Material

Plastics

The primary package in contact with the frozen product is generally a plastic derived from a natural hydrocarbon source. The choice of which plastic wrapper is dependant on the type of barrier required, and if the product is to be cooked or heated in the container. Migration of the plasticisers from the wrapping is a potential health hazard and the type of wrapping which can be in contact with food is covered by national legislation. The non-biodegradability of plastic wrapping material is an environmental issue and toxic compounds eg dioxins can be produced when, for example polyvinylidene chloride (PVDC) or polyvinyl chloride (PVC) plastics are incinerated at low temperatures.

The group of plastics called polyolefins, which include polyethylene, polypropylene and its co-polymers are most widely used. Generally, the higher the density of the polyethylene the better the barrier properties, with polypropylene being the best. Polypropylene is able to withstand temperatures up to around 100°C and is therefore suitable for boil in the bag products. To package ovenable products, a modified polyethylene CPET (Crystaline Polyethylene Tetraphthalate) is used, which maintains low temperature flexibility.

Polypropylene or polyethylene laminated with polyamide or polyester are frequently used for boil in the bag type products, and laminated plastic aluminium foil may be used when good vapour and moisture barriers are required, particularly with fatty fish to prevent fatty acid oxidation.

Plastic and paper based packaging is transparent to microwaves (passive packaging) and therefore can be used to contain foods which are to be microwaved. Thin polyester films can be metalised with aluminium and then laminated to a supporting board. When microwaved the aluminium absorbs a certain amount of radiation, generating heat and cooking the product.

Cartons

Cartons can also be regarded as primary packages when used as a protective sleeve to the product. The boards for the cartons can be made of:

1. Kraft boards. These are frequently used for packaging frozen foods and are usually made from fully bleached materials. They are strong, of good appearance and are suitable for direct contact with food.
2. Folding box boards. These usually have one fully bleached side which is suitable for direct contact with food.
3. Recycled fibre boards. These are usually used for secondary and tertiary packaging.

Secondary Packaging

Secondary packaging is usually a carton which holds a number of primary packages. The secondary package is usually made from boards but can be bands of paper or plastic.

Tertiary Packaging

Tertiary packaging is used to hold a number of secondary packages. Tertiary packaging may be palletised for easy handling and wrapped with shrink, stretch wrap or corrugated outers or may be packed in re-usable containers. Wooden pallets are in regular use, but can become a source of contamination. Plastic pallets, which can be colour coded, are more easily cleaned but will support the growth of mould in frozen fish factories.

Packaging Machinery

Packaging equipment may vary between a simple hand operated tool to an extremely complex machine and examples are listed in Table 15. For primary packaging of frozen fish two types of carbon, top load and end loads are generally used.

Top Load Cartons

Top load cartons which are formed immediately prior to packaging, arc supplied flat and unglued. The cartons are erected by a machine at speeds of between 20 and 320 cartons per minute and glued with hot melt adhesives which can tolerate cold store conditions. Top load cartons are used for some forms of fish sticks, the product being loaded in through the widest opening. A top load processing line normally consists of three machines for carton erecting, filling and closing respectively. Closing is usually effected by heat seal or adhesive to make the container tamper proof.

End Load Machines

End load cartons are used for products which are sufficiently robust to allow the product to be pushed in at the end of the carton. These are frequently used as a primary package for frozen fish portions or fish sticks. The process is slightly more expensive than top load processing, but has the advantage that the process is carried out by a single machine.

Auto Loading

Many machines have intelligent product - transfer units (IPTU's) which will automatically monitor and load the frozen product into the container and can be set to accept different tolerances. This technology is widely used in packaging frozen fish products to ensure that the weight of the product within the packet is with the designated tolerance.

Bags

Frozen fillets can be packed directly into bags made from materials with good gas vapour and moisture barrier properties. The level of sophistication can range from manual weighing and loading to a highly sophisticated form-fill-seal technology where a specified weight, volume or count of product is filled into a formed bag which is heat sealed. Such equipment can be used for packing IQF cooked, peeled shrimp.

In-Line Pouch Forming Equipment

In-line pouch forming equipment was developed in the mid 1960s to reduce the labour required to product the package, fill and seal the container. Such machines are widely used in the frozen fish industry, particularly in situations where sauces are added to the product. The material for the package is formed from a roll of packaging laminate (bottom web) which is heat formed in a dye, either by compressed air or vacuum to form the pouch. The product is then loaded, either by machine or hand into the pouch and any added liquid will be added by an automatic depositor. The package then has the top sealing web brought into contact with the bottom web, evacuated and then heat sealed. Some products may have an inert gas injected into the package as an alternative to evacuation, to prevent the packaging coming into contact with the product. The packages are then separated by cutters which cut both along and across the webs which join the packages together.

Machinery For Tertiary Wrapping

Tertiary packaging is the final stage of the packaging procedures and machinery is available which will collate, package and palletise the secondary packages. Tertiary packaging normally takes place after the product has been frozen. A typical operation would involve the orientation and collation of a number of secondary packages which would be automatically loosely wrapped in shrink wrap. The shrink wrap would then be heat sealed before the wrap packs are conveyed through a heated tunnel, where the wrapping shrinks to assume the geometry of the pack. Alternatively, the secondary packages can be packaged in cases, which may be formed round the product (wrap around) or the case may be preformed and glued before the packages are loaded.

The tertiary packs are then frequently palletised for storage or distribution. The pallet and contents themselves may be wrapped in stretch wrap or similar materials for further protection. This may be done by hand, but is increasingly carried out by mechanical pallet wrappers.

Cold Stores

A cold store is any building or part of a building used for storage at temperatures controlled by refrigeration at -1WIC or lower.

Recommended Storage Temperature

The spoilage of fish due to protein denaturation, fat changes and dehydration can all be slowed down by reducing the storage temperature.

The recommended storage temperature for all fishery products in the UK is -30°C and this temperature has also been adopted throughout Europe. Spoilage by bacterial action in any practical sense is completely arrested at this temperature and the rate at which other undesirable changes proceed is greatly reduced. Some products can be stored safely at higher temperatures than the -30°C recommended providing storage is only for a short period. Since it is not always possible to guarantee that a product will stay in storage no longer than originally intended, it is generally safer to use the lower recommended temperature.

The International Institute of Refrigeration recommends a storage temperature of -18°C for lean fish such as cod and haddock and -24°C for fatty species such as herring and mackerel. The code also recommends that for lean fish intended to be kept in cold storage for over a year, the storage temperature should be -30°C.

Cold store operators can seldom be sure to store only one species or type of fish, or to store it for a limited period only. Cold stores built for storage of fish should preferably be able to operate at -30°C but can be operated at a higher temperature if circumstances and relevant codes or recommendations allow.

It has been calculated by an eminent authority on cold store design that under specific conditions, the total cost of operating a cold store at -30°C is only 4 percent higher than when operating at -20°C although the corresponding percentage increase in running costs will be higher.

Factors Limiting Storage Life

Protein changes. Fish proteins become permanently changed during freezing and cold storage and the speed at which this denaturation occurs depends very largely upon temperature. At temperatures not very far below freezing point, -2°C for example, serious changes occur rapidly; even at -10°C, the changes are so rapid than an initially good quality product can be spoilt within a few weeks. The rate of deterioration due to protein denaturation, however, can be slowed by ensuring that storage is at as low a temperature as possible.

Fat changes. Fatty fish may become unpleasantly altered during cold storage but they can be protected to some extent either by glazing or by packaging in plastic bags sealed under vacuum. These oxidation changes take place more rapidly at higher temperatures and storage at a low temperature is an effective means of slowing the rate of spoilage by this method.

Colour changes. The quality of fish is often judged by appearance, and colour changes which are not otherwise significant can result in fish being downgraded. The changes in the fish flesh which bring about these colour changes are also retarded at lower temperatures.

Dehydration changes. Dehydration of the product is probably the major concern of the cold store operator and the rate of drying can be linked with a number of factors in cold store design and operation.

When fish get badly dehydrated in cold storage, the surface becomes dry, opaque and spongy. As time progresses, these conditions penetrate deeper into the fish until it becomes a fibrous, very light material. Visible effects of severe dehydration on the surface of the fish are known by the term "freezer burn". This is an unfortunate choice of term since the effect is unlikely to result from freezing in a properly designed freezer, and appears only after periods of storage in a cold store.

Frozen fish may dry slowly in cold storage even under good operating conditions. This is undesirable for reasons other than the obvious one that the product will lose weight. Drying also accelerates denaturation of the protein and oxidation of the fat in the fish. Even totally impervious wrappers used to protect the product do not give full protection if the cold store operating conditions are favourable for desiccation within the pack. In-pack desiccation prevails when there is some free space within the wrapper and the temperature of the store fluctuates. When this occurs, there will be times when the wrapper is colder than the fish and moisture will then leave the product and appear as frost on the inner surface of the wrapper. The total weight of the product and package will not change but if the in-pack dehydration is severe, the fish will have the quality defects of excessive drying.

Choice of Planners and Designers

When considering the construction of a cold store, one of the very first steps is to decide on specialists to be responsible for planning, design and project management. The construction of cold stores involves a number of factors besides the actual building. The use of specialists enables the organisation responsible for the project to:

- Share the responsibility for the project with an outside body.
- Avoid building up a department of costly specialists who may not have adequate knowledge in the field of activities and who eventually will not be fully occupied with the project.
- Benefit from the practical experience of the specialist group.
- Save time as initial training and research will not be required.
- Ensure that latest techniques are used.

It is recommended that specialists handle a project all the way from a feasibility study to commissioning, including supervision and training of the local management responsible for the future operation of the cold store.

Shape and Size

Cold stores can be divided according to construction into single-storey and multi-storey buildings. They can be used as production stores, bulk stores, distribution stores or retail stores. For a long time, the most appropriate shape was a cube for which the ratio of surface to volume is a minimum. Besides this, the cost of land was a major consideration, especially when stores were located in urban areas. This resulted in multi-storey buildings, with a number of disadvantages, e.g., costly foundations, heavy framework, congested handling areas.

The main considerations which have resulted in the appearance and success of single storey buildings are cost reduction together with mechanised handling techniques. Today multi-storey buildings are built only in congested or costly harbour areas, where cranes can be used externally to mezzanine floors. Those buildings are normally not more than two storeys high.

A single-storey cold store can be easily designed to meet the specific requirements of stacking and handling equipment. Wall and roof constructions can be made lighter as they do not have to support the weight of the product stored, as in a multi-storey building. The main disadvantages are the relatively large ground area covered and the high ratio of surface to volume. The advantages however, normally override the disadvantages. Most European and United States cold stores built in the course of the last 20 years are single-storey buildings.

A production cold store is usually a part of one or several food businesses, storing frozen raw materials and semi-finished, as well as finished products. Bulk stores normally give the same service as production cold stores, but are often located at some distance from the actual processing industries and are normally much larger than the production stores. The storage time at a bulk store is also normally longer. Distribution stores, generally located in urban areas, receive products from the production or bulk store in large lots, which are broken down - order assembled - before delivery to the retail stores. The storage time is short, one week up to two months.

There is a general tendency to build larger installations than those in the past. Capacities of new cold stores are now between 5,000m^3 and 250,000m^3 where, depending on local conditions, the optimum investment/running cost relationship is generally found. It should, however, be noted that the size of the distribution warehouse depends on a number of factors like amount of traffic, average storage period, number of articles, as well as the number of clients. Bulk and distribution cold stores are often combined at the same location, the main difference between them being stacking arrangements and equipment used.

General Layout

A single-storey building can have a relatively simple layout. Depending on size, it can either have one single room or it can be divided into a number of rooms. Normally all the rooms are operated at the same temperature, for fish preferably in the range of -24° to -30°C. Most stores, with the exception of small prefabricated ones, are built at a higher level than the surrounding yard with a special loading ramp at one or more sides. The loading ramp level corresponds to the height of the most commonly used vehicles. Sometimes the stores are also built with a loading ramp for railway wagons, often placed on the opposite side to that used for loading road vehicles.

The engine room should be as close as possible to the position of the air cooling equipment within the store. This sometimes poses a problem in the planning of future extensions and it may therefore be placed at the end of the cold store in such a position that it will easily serve future expansion. Alternatively, the engine room can be placed away from the cold store complex and serves the air cooler via a pipe bridge so that extensions can take place in any direction. Freezing tunnels can either be arranged so that their entrance doors communicate with the loading platform or alternatively they can be arranged with the discharge doors within a cold room so as to minimise the amount of heat losses.

Construction Methods

Modern large or medium cold stores are built as one-storey buildings designed for mechanical handling, e.g., forklift trucks and automatic stacker cranes. Manual handling is, however, still used for most small-sized stores.

A cold store can be built as an ordinary building using conventional building material, such as bricks, concrete or concrete process sections to which a vapour barrier and insulation is fitted internally. Modern insulation material, in particular polyurethane, has a strength that can be utilised structurally. Today, this is used for panel designs suitable for all sizes of cold rooms from. Factory made insulation panels are delivered to the site complete with a vapour barrier and internal cladding, thus reducing the site work to a minimum. There are two basic principles for panel-built cold stores. A common system has an external structure and cladding with a wall insulation on the inside of the columns and the insulated ceiling hanging from the outer roof structure.

The panels normally used in these systems are either polyurethane or polystyrene insulated panels with or without frames. They are manufactured as sandwich panels, one face being the vapour barrier of light-gauge galvanised steel sheet and the other face being the internal finish of plastic-coated galvanised sheet or aluminium sheet. A decorative external cladding is erected on the outside of the columns.

The roof insulation is constructed as a suspended ceiling. The roof panels are, in principle, the same as the wall panels, but are sometimes equipped with wooden frames.

The wall panels are fixed to the columns or horizontal rails between the columns with special bolts. The joints are sealed with tape or sealant mastic and the internal joints are finished with a cover strip. The roof panels are hung from the outer roof construction with hanger rods and locked together with sealed, tongued and grooved joints or similar. For the roof panels, special care must be taken where the hanger rods pass through the vapour barrier. In humid climates, ventilation might not be sufficient to avoid condensation in the attic space above the insulated ceiling. This problem can be overcome by closing the space and drying the air with some form of air drier.

The latest development are panels with a rib profile on the external face of aluminium which is also the total external cladding - with polyurethane foam insulation and an internal face of low profile corrugated aluminium. The panels are generally of large size and erected with small mobile cranes. Thus the erection time is reduced to a minimum, the panels are pulled together with Camlocs or other special devices giving positive pressure between the joints. The joints are filled with flexible compounds and covered with a metal strip. The walls are attached to horizontal rails fixed at 3 m vertical intervals with clamps, which admit sufficient movement of the construction. These panels are also used for roofing, being place on beams at 3 m intervals. The external cladding joints are made with a special seaming machine, which automatically moves along the joints, mechanically closing the rib profile of one sheet around that of another. The space between the panels is filled with a one component polyurethane foam. The internal cladding is sealed with a PVC strip. One component foam is also used to join wall panels to roof panels and seal around doors, etc, maintaining good insulation properties throughout the building. With this design both the insulation and external vapour barrier are entirely sealed units enveloping the whole building. This means that losses via heat bridges or air leakage are completely eliminated, which gives practical insulation properties closer to theoretical values than normally expected.

In the case of single-storey cold stores, two types of frame work are commonly employed. Metal frameworks, can span distances up to 60m without the need for internal columns. They are prefabricated in the factory and transported to the site in sections for quick and easy erection. The minimum span is approximately 15m. Minimum load is involved as the roof frame is carrying only the waterproof covering and the insulation. In some designs it also carries the weight of the air coolers within the room. Then it is desirable to concentrate these loads near the columns rather than at mid spans. An outside metal framework can be used for electrical earthing connections.

Reinforced concrete frameworks can incorporate concrete beams spanning the room, or can be a combination of concrete columns supporting metal trusses. Overhead rail systems can be supported from the main structure or a separate steel frame can be incorporated inside the cold store with separate columns transferring the load onto the main structural floor.

Insulation

The choice of insulation is very important as it accounts for a large proportion of the total construction cost. The insulation material and thickness is also important from an energy point of view. Besides a satisfactory thermal conductivity coefficient the insulation material should also be odour-free, anti-rot, vermin and fire-resistant and impermeable to water vapour.

Currently, with existing energy costs, the thermal conductance should not exceed 0. 15 kcal/m^{2}h°C for cold stores. However in the future with ever increasing energy costs this figure may have to be improved.

The final quality of any insulation is not only a matter of the properties of the material itself, but of the way it is erected or fitted to the external building. Heat bridges should be avoided, e.g., those normally created by pipes, cable joints, etc. Piping which carries low pressure refrigerant or other liquids at low temperature must be insulated. The provision of an efficient vapour barrier on the outside of the finished insulation with joints properly sealed is of utmost importance, as moisture vapour penetrating the insulation will form ice and gradually destroy the insulation material. The thickness of insulation depends upon the internal temperature, heat conductivity of the insulation material and the dew point of the ambient air, in order to avoid condensation. The insulation material should be protected against moisture and mechanical damage. Where uncovered insulation material is used, the internal walls and ceiling can be protected by sheets of aluminium, galvanised steel, reinforced plastic, etc., or with materials such as plaster and cement. The choice of material should be related to the use of the store, e.g., need for washing down. Painting of plastered walls are not recommended unless special paint is used as it will quickly peel off.

The insulation of the cold store doors should be the same standard on the store wall. The most common insulation material for doors is polyurethane and door heaters should be fitted to prevent ice forming at the seal thus jamming, and ultimately causing damage to the door.

Vapour Barriers

The air within a cold store holds a good deal less water vapour than the air outside. Water vapour in the air gives rise to a pressure and together with the other gases present, such as oxygen and nitrogen, account for the atmospheric pressure that we are all familiar with. The partial pressure exerted by the water vapour is proportional to the quantity of vapour present and the vapour in the air will tend to migrate from areas of high partial pressure to areas of low partial pressure. Hence, there is a tendency for moisture in the ambient air to pass through the insulation of a cold store to the area of low partial pressure within. When this vapour is cooled, it condenses and at the point where the temperature is 0°C, it freezes to form ice. This process will continue over a long period of time and the build-up of ice will eventually affect the insulation properties of the cold store wall and also weaken the structure of the wall or building. Unfortunately, the outward effects of this build-up of ice may not show for some time, long after the builder's guarantees have become invalid.

To prevent this type of destruction to the store insulation, a vapour barrier has to be provided on the warm side of the insulation. This vapour barrier must be complete and cover all walls, the roof, ceiling and the floors. For stores constructed against a building wall, this may be formed by applying at least two coats of a suitable bituminous sealing compound. With prefabricated stores, a vapour barrier is already provided with the individual sections, usually an outer facing of sheet metal, and only the joints require sealing. It must be remembered that water vapour is a gas and it is not sufficient merely to make the outer surface waterproof; overlapped joints, for instance, must be sealed.

Foundations and frost heave. Low temperature stores built directly on the ground may require special precautions to prevent the build-up of ice below the cold store floor. The ice formation causes distortion known as "frost heave" and in particularly bad cases, it can lead to the complete destruction of the store and structure of the building.

The conditions that give rise to frost heave are rather complex, since they are related to the type and texture of the soil, the insulation properties, the availability of moisture, the dimensions of the store, seasonal climatic variations and other factors.

Two methods of preventing frost heave are commonly used. The ground below the store can be heated either by a low voltage electrical mat in the cold store foundation or by circulating a heated liquid such as glycol through a pipe grid built into the foundation. The heat for the glycol is usually obtained from the compressor hot gas through a heat exchanger.

Another method of preventing frost heave is to leave a space below the store for ventilation. The level of the floor of a cold store is usually arranged to suit the unloading and loading of vehicles. The additional height required for this facility leaves plenty of height for an air ventilation space below the insulation. If there is any danger of flooding, cold store floors will be built above the likely water level and again there will be an opportunity to leave an air space for ventilation. This ventilation arrangement should be clearly defined and not blocked at a later date when the main function of the air space has long been forgotten.

*Air ingress.*Outside air entering the store adds heat and moisture. This moisture will be deposited as frost on any cold surface and will eventually finish up on the surface of the cooler. Excessive air exchange should be prevented to keep the cold store temperature steady and reduce the frequency of defrosting. Small air locks have been used to prevent the free flow of air in and out of the store but they are not popular with cold store operators. The air-lock space often does not allow complete mobility and unless this condition can be met, both doors are left open. The air lock will therefore serve no useful purpose and merely occupies valuable space.

A curtain of air blown downward or from the side of the doorway can reduce the exchange of air when the door is open. These air curtains, as they are called, can be a useful aid when the door is opened for short intervals. However, they are often abused and doors are often left open for long periods.

Hatches can be used to reduce air ingress when a product is being loaded or unloaded. Hatch openings should be as high up in the store wall as possible to prevent excessive loss of cold air. Portable conveyors can also be used to speed up the transfer of produce.

Store door openings can be fitted with an inner curtain made from overlapping strips of synthetic material suitable for use at low temperatures. This reduces the air exchange considerably without interfering too much with traffic but the curtain must be maintained in good order and, as with the air curtain, not abused by leaving the outer main door open.

Large stores are fitted with power-operated doors which can be quickly opened and closed, usually by automatic vehicle sensor or pendant switches outside and inside the doorway. Because this system is easy to operate even from a moving fork lift truck, door opening times are kept to a minimum.

Floors

The ground loads from a cold store are in the order of 5500-8000 kg/m^2. This consists of static loads due to merchandise, structure and concentrated rolling loads transmitted by e.g., forklift trucks and other handling equipment. It is of importance that those loads are investigated in detail for each special project.

In the case of a single-storey building, a reinforced raft is usual, including ground beams at the edges or bases for the structural frame. This can rest directly on the existing ground or a supported slab.

The floor wearing surface requires particular care. In addition to the wear other industrial floors have to stand, it is exposed to low temperature. All other parts of the cold store can be repaired whilst most of the space is still used for storage, but not the floor. Most commonly the floor wearing surface is a concrete slab cast on the floor insulation with a thickness of 100-150mm. In cases where intensive traffic is foreseen a special hard wearing top-finish is recommended. Before casting the wearing surface, the floor insulation should be protected by bituminous paper or plastic sheeting, the function of which is twofold. Firstly, to prevent the water from the fresh concrete penetrating into the floor insulation and secondly, to provide a slip-sheet, which will reduce the friction when the concrete when contracts. It is of great importance that the floor wearing surface be level to enable high stacking and easy traffic. The top-finish should provide a reasonable anti-slip surface.

Special attention must be given to floor joints. It is recommended that a device which allows horizontal displacement, but not vertical movement, is used between the joints. If the joints open too much after lowering of the temperature, they must be filled with a suitable jointing compound.

Types of Cold Stores

Stores with unit coolers. The most widely used method of cooling modern cold stores is by means of unit coolers with fan designed with good air flow characteristics (or good circulation of the air). This type of cooler is generally the cheapest to install; it contains a relatively small charge of refrigerant, it can be readily defrosted without interfering too much with the store conditions and it does not require a heavy structure for support. The main disadvantage is that many designs using this type of cooling unit do not allow for uniform distribution of the air within the store. This gives rise to poor storage conditions where the air circulation is either too high or too low. By suspending the unit cooler from the ceiling) or installing the unit outside the store and ensuring that pallets are stacked with suitable head space and floor spacing, uniform air distribution can be achieved.

Multiple units are usually better than large single units for a number of reasons. A multi-unit system gives some insurance in case of breakdown. The store can usually be maintained at its design value without the need for all units to be in operation provided there is not a high additional refrigeration load due to product and heavy traffic in and out of the store.

Multiple units also allow each unit to be defrosted in sequence and this arrangement has the least effect on storage conditions. If a hot gas defrost system is used, then a multiple unit system is essential so that the units in use provide the necessary refrigeration load for the refrigeration compressor.

With small units, electrical defrosting is more common. The defrosting of unit coolers in small cold stores is usually automatic and operated by a time clock. With this mode of operation, the timing of defrosts should be arranged to coincide with times when the refrigeration load is low, usually during the night.

Prefabricated Cold Stores

Besides prefabricated panels and the structural components used in the construction of cold stores, there are "building kits" available on the market today for small modular cold stores. The most complete "kits" include wall and roof panels, loading ramp, canopy as well as refrigeration plant. A typical example is a cold store with a nominal storage capacity of some 200t measuring 12 x 12 x 6m built with self-supporting polyurethane insulated panels faced inside and out with galvanised and plastic coated steel sheeting, as well as a prefabricated floor. The only local requirement is a concrete floor slab on which the building is erected. Normally the assembly is carried out by specialists and the erection time varies between 4 and 8 weeks depending on local conditions. The material for the store is shipped in three ordinary containers one of which contains the engine room which can be contained in a weatherproof building adjacent to the cold store.

Cold-Air Distribution

Heat transfer is effected by radiation or convection. The air in a cold room essentially transfers heat by convection. Convection is often referred to as natural when air movement is activated only by a density difference created by the temperature difference. It is called forced when air movement is activated by a fan. The actual cooling is effected by two main types of heat exchangers, natural convection coils and forced-air coolers. Natural convection coils have the advantages of maintaining high relative humidity and low air velocities, but these advantages are offset by disadvantages like difficulties of defrosting, which also could be dangerous to carry out. Furthermore, they are not suited to run with high product loadings since the refrigeration capacity is low and are therefore rarely installed today.

Forced air coolers can be mounted inside the cold store space or placed in an external compartment with air circulation by means of fans through a delivery duct. Such a duct sometimes taking the form of a double ceiling or double floor. The defrosting of air coolers situated externally is convenient since it is possible to isolate the air cooler from the interior of the room for this operation. The normal placement is, however, inside the room for larger cold stores, whereas outside placement is normally used for smaller ones. Forced-air circulation enables greater refrigeration capacities because of high rate of heat transfer. It also gives a more even temperature distribution within the room.Forced-air coolers are usually built as a single small unit, including the fan which is easily mounted within the room itself. The equipment is often combined with special air ducts for even air distribution in the room. The advantages of this type of equipment are reduced installation costs and easy maintenance.

Defrosting

When the refrigerant temperature is lower than -3°C, frost will deposit on the coils and this results in a reduction in the heat exchange. The frost thickness built up is however, of less importance than ensuring the free passage of air through the coil battery as indicated above. Regular defrosting is of great importance in the operation of a cold store. There are a number of methods available, such as hot-gas defrosting for direct expansion systems, water defrosting and electric defrosting. Sometimes combinations of these methods are employed, e.g., hot-gas defrosting followed by water spraying or hot-gas defrosting of coils with electric tray heating. The latter is now the most used in new installations. It should be noted that labour costs for manual defrosting operations can be high and they are often complicated. In order to increase the operation periods of air coolers between defrost, a wide fin spacing of coils at the inlet side is used in order to act as frost catchers without obstructing the airflow.

Factors Affecting Storage Conditions

The rate of product dehydration can be related to the size and shape of a cold store. A small cold store has a greater heat leak in proportion to the quantity of product in the store since the volume of a store increases at a greater rate than the surface area. This means that one large store is likely to provide better storage conditions than two smaller stores with the same capacity.

Product Handling and Storage

Large stores are provided with a loading platform which can be adjusted to accommodate varying vehicle heights. This platform must also provide adequate space for quick sorting and manoeuvring of goods in and out of storage. A platform width of 8 to 10m may be necessary for this purpose. The unloading area should also be roofed over so that goods being transferred in and out of the store are protected from direct sunlight and rain. This cover also protects the doorway, which may ice up if it is exposed to rainfall.

In hot countries, handling frozen fish outside the low temperature storage space can quickly result in exposed fish being warmed and even thawed. The provision of a refrigerated working area and loading dock is therefore recommended for prestorage sorting and the assembly of loads for shipment. This loading dock should be totally enclosed, insulated and refrigerated to a temperature of about MC. The area of this dock will depend on the amount of traffic and the type of store operation.

In addition to providing a chilled working space, this refrigerated dock will act as a large air lock between the outside air and the low temperature air within the store. As much as 80 percent of the moisture in the ambient air will be removed by the cooler in this space and a good deal of precooling will be done before this air enters the main store. This will reduce the defrost requirement for the store coolers and generally result in a more stable and lower storage temperature.

The means of transporting goods in and out of the store and within the store depends on the goods being handled, the type of cold store, the height of store, the need to reduce labour costs and many other factors that may only have a local significance.

Whenever possible, pallets should be used for storage of product. These divide the goods into unit loads which can be transported, stacked and retrieved with a minimum of effort. Regular-shaped packages or blocks can be readily palletised. Loose fish, such as those broken from blocks and other irregular- shaped products, can also be stored in pallet cages. In public stores where it is often necessary to remove a pallet from the bottom of a stack, the individual pallets do not rest on the pallets below but are supported on a framework. This allows any individual pallet to be added or taken away without the need to break down the stack.

Pallets should not be stacked so that the base of one pallet rests on the produce below except in the case of frozen blocks of fish or where the product cannot be crushed. Framed pallets can be stacked five high with safety, but only if they are correctly stacked. In large distribution stores, pallet racks have been motorised so that there is no need to provide so many passages within the store. The racks are moved as required to allow access to individual rows. This degree of mechanisation would only be employed when store utilisation and quick handling are critical factors.

Attempts have been made to standardise pallet sizes but this has not yet become world-wide. Pallet dimensions of 800 x 1200mm and 1000 x 1200mm have been widely used but the final choice will depend on local circumstances depending on such factors as the degree of interchange of pallets outside the store, vehicle and package dimensions, and other transport and storage considerations.

When a fully accessible palletised system is not used, the product should be loaded in the store so that a first-in first-out system can be operated. This ensures that there is a correct product rotation, and storage times are not unnecessarily long.

The width of passageways will depend on the equipment used for transporting and stacking the product. Details of the space requirements of this equipment must therefore be obtained before a decision is made on the size of store required.

When products are placed in the cold room it is important that an air space is left between the product and the ceiling, the floor and the external walls otherwise heat entering the store through the insulation will pass through the produce before being transferred to the cooler. In the case of internal walls an exception can be made only when the same temperature exists on opposite sides of the wall.

With normal storage of palletised products the required air space is usually obtained through the small irregularities which occur when assembling the product on the pallet. However, in the case of solid block storage or where the pallet sides are completely flat, special care should be taken to ensure that the air space is adequate. Between the product and the floor the air space is automatically provided by the construction of the pallet. The question of air space above the uppermost pallet is, as a rule, no problem since the height of the chamber is designed for a certain number of standardised pallet units and thus allowance is made at the design stage.

Refrigeration

The capacity of the refrigeration plant must be based on a thorough heat load calculation for each individual project. Refrigeration load can vary widely for stores of the same capacity depending on design, local conditions, product mix, etc. Therefore no rule of thumb can be applied. In past practice, a safety margin of some 50 percent of the theoretical calculation has been used. Today with a more thorough knowledge of practical cold store operation, combined with theoretical knowledge, the safety margin can be reduced to a more realistic level. The refrigeration equipment should conform to requirements laid down in national codes of practice, insurance companies, as well as international recommendations

Heat leakage or transmission load can be calculated fairly closely using the known over-all heat transfer coefficient of various portions on the insulated enclosure, the area of each portion and the temperature difference between the cold room temperature and the highest average air temperature likely to be experienced over a few consecutive days.

Heat infiltration load varies greatly with the size of the room, number of door openings, protection of door openings, traffic through the doors, cold and warm air temperatures and humidities. The best basis for this calculation is experience. The type of store has a marked influence on the heat load, as has the average storage time. In comparing long-term storage, short-term storage and distribution operation it can be found that there is a 15 percent increase in refrigeration load for the short-term storage as compared to the long-term storage, whereas the refrigeration load in the distribution operations is in the order of 40 percent higher than for long-term storage, due mainly to additional air exchanges.

Most large cold stores are equipped with 2-stage ammonia refrigeration installations. For smaller plants, usually less than 6 000 kcal/h refrigeration capacity, approved refrigerant will probably be used in single stage systems operating with thermostatic expansion valves. Such systems are thermodynamically less efficient, but in areas where only staff with relevant refrigerant experience are available the system may be preferred for service reasons.

The refrigeration system should be designed for high reliability, and easy and proper maintenance. Once a cold store plant has been pulled down in temperature, it is expected to maintain this temperature, literally, forever.

Even maintenance jobs that need carrying out only every 5-10 years must be taken into consideration.

Cold Store Capacity

There is no method of defining cold store capacity that satisfies the requirements of everyone concerned with cold storage. Storage capacity based on the weight of produce that can be stored will depend on the storage density of the products and the method of storage.

Therefore, unless only one product is stored under closely defined conditions, this definition is obviously unsuitable. It is generally agreed that it is more appropriate to define storage capacity in terms of the store volume but there are a number of ways in expressing this value.

Gross volume is the volume of the refrigerated space.

Net volume is the volume that can potentially be used for storage and is the gross volume less the volume required for coolers, structural requirements, doorways and other permanent features of the store.

Effective volume is the store space that can actually be utilised for storage and it takes into account the requirements for passageways, stacking equipment etc.

Gross volume and net volume can easily be defined by devising a simple set of rules for making these calculations. These store volumes, however, can only give a rough estimate of storage capacity and their main use may before statistical purposes. The effective volume can only be calculated for each particular case and to achieve any degree of accuracy, a drawing of the store layout would be required together with full details of the storage conditions. Store operators should therefore use general statements of store capacity with care and when placing an order they would give full details of the products and the storage operation to enable the supplier to provide a store to suit the operating requirements with the maximum utilisation of the gross storage volume.

Freezer Weight Loss

Weight may be lost by dehydration or due to physical damage of the fish during the freezing process.

Physical damage may be due to damage during freezing which results in small pieces being broken off; this is likely, for instance, in freezers where the product is fluidized by the cooling air.The other form of physical damage encountered during the freezing process is due to fish adhering to trays or conveyor belts. If the weight loss on releasing fish from trays is excessive, the trays may be sprayed on the underside with water to assist release. Fish frozen in continuous freezers with stainless steel link or mesh belts may stiffer weight losses due to small particles being trapped in the belt. Losses due to physical damage in a freezer should be small and need not be more than about 1 percent if the freezer and freezing process is suitable for the product.

Weight loss due to dehydration in a freezer depends on a number of factors, and the weight losses in air blast freezers give rise to the greatest controversy. Weight loss due to dehydration will depend on:

— Type of freezer
— Freezing time
— Type of product
— Air velocity
— Freezer operating conditions

Freezers such as plate freezers where the fish is frozen by contact and released by defrosting will have a negligible weight loss during freezing. Any measured change in weight will probably be due to loss of drip before the freezing started.

Dehydration losses occur mainly in air blast freezers and in other freezers which use a gas such as nitrogen or carbon dioxide in direct contact with the product.The loss of weight in nitrogen, carbon dioxide and other cryogenic freezers will be low by virtue of the fact that freezing times are short. A direct contrast made between a carbon dioxide freezer and an air blast freezer showed that the weight lost from haddock fillets in carbon dioxide freezer was about half of the weight lost in the air blast freezer, 0.6 percent compared with 1.2 percent. Other cryogenic freezers are likely to give rise to weight losses which are about the same as that of the carbon dioxide freezer.

Cold Store Weight Loss

Much has yet to be done to correlate the rate of weight loss with differences between storage conditions but the rate of weight loss has been shown to vary with the following:

— Temperature
— Temperature fluctuation
— Humidity
— Heat transfer
— Air flow over the product
— Radiation effects of lighting
— The product
— Shape and size of the product
— Type of wrapper

Most codes of practice only state the temperature for storage. Variations in the other factors that control the rate of dehydration can therefore result in cold stores having widely different storage conditions. The rate at which the product loses weight by dehydration can therefore vary considerably.

The losses are expressed as the weight changes per square metre of exposed fish surface. These results clearly show that there are great differences between the quality of cold stores which may be attributed both to their design and mode of operation as well as to the operating temperature. Apart from the physical loss in weight, excessive dehydration results in "freezer burn". The overall weight loss, however, cannot be used to define the point when "freezer burn" becomes apparent. Dehydration only occurs from exposed surfaces and the rate of dehydration is greater where the surface area to volume ratio is high. The edge of fish fillets and the corners of slabs of fish will therefore show signs of excessive dehydration or "freezer burn" long before the other exposed surfaces of the product. For this reason, "freezer burn" can even become apparent on glazed fish long before the overall weight loss is equal to the weight of glaze applied.

The rate of weight loss within a store can vary considerably with location. Fish stored near fan coolers, where they are subjected to high air velocities, will quickly show signs of dehydration. Fish stored against walls remote from the cooler may be subject to poor air distribution and heat gains from the store walls. This can cause temperature fluctuations in the product which inevitably results in high dehydration losses.

Fish adjacent to roof coolers may also dehydrate more quickly since the path of moisture migration is considerably shorter. Fish near the roof or walls of stores which are affected by a high incidence of solar radiation may also be subjected to higher losses. Finally, fish in storage which have unfrozen or partially frozen fish frequently stacked alongside show the highest dehydration rates of all.

Freeze at Sea

The length of time a fishing boat can remain at sea depends on the time the fish can be kept so that they are still edible on reaching the consumer. Storage in ice or by other means which keep the fish chilled is adequate for periods not much in excess of two weeks.

Fish such as haddock and cod caught in the North Atlantic fisheries can be stored for up to 15 days in ice and then rapidly become inedible. It has been found that fish caught in tropical waters can remain edible for even longer periods when stored at chill temperatures. This may not be a general rule and the limitation of chilled storage must be established by local experience.

In practice, the time restriction for storage in ice often means that fishing vessels must return to their home port with the fish room partly empty. There is therefore a need for some means of preservation that will extend the storage life without substantially altering the nature of the raw material. Quick freezing and cold storage is an excellent way of doing this.

When newly caught, fish are frozen quickly and stored at a low temperature on board, so there is no limit imposed on the length of voyage due to spoilage of the catch. Fishing vessels can remain at the fishing grounds until the hold is full. This increases the proportion of time spent at the fishing ground and improves the economics of fishing. It also allows the fish to be distributed to a wider market even without the existence of an elaborate "cold chain ". Fish which have been frozen at sea are of very good quality when landed; therefore, more time is available for the fish to be distributed over a wider area and still be in good condition.

Freezing at sea has therefore an important role in world fisheries. A look at a map will show that large areas of ocean are far distant from any centres of population or even land, and many potential fisheries are therefore unexploitable without a method of preserving the fish for long sea voyages. Only quick freezing and low temperature storage has so far satisfied this need and, as traditional near water fisheries become overfished or are unable to satisfy the growing demands of an ever increasing population, freezing at sea will become more and more necessary.

Type of Freezer Vessel

Fish frozen at sea may be frozen whole, immediately after catching and when thawed on shore, can then be used in much the same way as fish traditionally preserved in ice. Alternatively, the fishing vessel can operate as a fish processing factory and the fish may then be filleted, packaged and frozen, and the waste products converted to fish meal and oil.

Freezing of the whole fish has the following advantage over processing before freezing. The number of crew required is not much greater that for a fishing vessel of comparable size preserves its catch in ice. Processing equipment and factory deck space, are a good deal less. The whole fish, when thawed after landing, are available for any form of traditional processing. The problems associated with freezing newly caught fish are less with whole fish than fillets. For the above reasons, it may therefore be advisablc as a first step for a developing country to freeze whole fish and progress to a factory-freezer operation as the situation demands.

How Good. Are Sea-Frozen Fish?

Sea-frozen fish, properly handled between landing on deck and loading into the freezer, when thawed are almost undistinguishable, from fresh fish kept in ice for a few days.

The loss in quality as a result of freezing, cold storage and thawing is small when these treatments are properly applied. Thus, when very fresh fish are frozen at sea, the final product can be equal to the best on the market.

Freezers for Use at Sea

A number of conventional freezer units may be used at sea with little modification but may have to conform with national regulations and insurance requirements for fishing vessels. Many countries, for instance, do not allow the use of ammonia as refrigerant because of its toxicity and because there is a potential explosion hazard. The design and operation of the freezers and the refrigeration system must take into account the movement of the vessel, vibration, sea-water corrosion and the extra rough usage likely under the arduous conditions experienced at sea. Another factor that may influence the choice of type of freezer is the type and variety of fish species to be frozen. The freezer should be able to cope with the variation in fish sizes in fisheries where many different species are caught.

The VPF was specially designed for freezing whole fish at sea. In most applications, a 100mm spacing between plates was found to be adequate. This spacing allowed a very high percentage of the catch to be quick-frozen and reduced to the cold storage temperature of -30°C in the recommended time of 4h. Oversized fish were normally frozen in a separate air blast freezer room.

The UK design considerations which led to 100mm spacing being used, were based on the freezing of gutted fish with heads on. Other countries' requirements may be for ungutted fish to be frozen or for fish to have the heads removed as well as guts before freezing. These considerations will have to be taken into account as well as the size, shape and variety of species to be frozen before a decision is made on the preferred plate spacing.

Another factor to be taken into account with any type of freezer is the overall size and weight of the frozen product. If the product has to be lifted and stacked in the fishing vessel cold store, care should be taken that the block weight is well within the physical capabilities of the crew. It has been possible to operate in the UK with 45 to 50 kg blocks measuring approximately 1060 x 520mm.

The following design and operational requirements for freezers to be used at sea will give the reader guidance on whether a freezer is suitable for this application.

1. The freezer should be easily loaded and unloaded.
2. Freezers with trolleys should have special arrangements to make them safe during rough weather.
3. The freezer should be able to retain the product during the loading and unloading procedures; serious damage or injury can result from a dislodged frozen block or tray of fish.
4. The freezer should be able to operate with part loads which may result from variations in the catching rate.
5. The refrigeration system should not give rise to uneven freezing due to displacement of the refrigerant with the movement of the vessel.
6. The freezer should be robust.
7. The material used in the construction of the freezer should be resistant to seawater corrosion.
8. The freezer should be constructed so that it can be cleaned by hosing with seawater.

The above list is not exhaustive but it is sufficient to indicate that many types of freezer would, in fact be unsuitable for use on a fishing vessel.

Space on a fishing vessel is limited especially the height available between decks. Freezer designs and layout should therefore be made to suit this space restriction. The quicker the freeze, the smaller is the size of a freezer for a given capacity. Freezers for use at sea should therefore be designed for short freezing times, taking into account both the refrigerant operating conditions and the product shape and size. Large fish like tuna are frozen individually. Brine immersion freezers have been used but there has been a recent trend toward air blast freezers for this purpose. Shell-on shrimp and other shellfish are also frozen individually, but apart from these few exceptions, freezers for IQF products are unlikely to be required on a fishing vessel.

Extra care has to be taken with the refrigeration equipment. Pipework should be secure and routed so that it is unlikely to be damaged. The use of secondary refrigerants for many shipboard systems has resulted in a system that can be maintained in a relatively leakfree condition. When a secondary refrigerant is used, the primary refrigerant is confined to the condensing unit and heat exchanger. Since secondary refrigerants are liquids at atmospheric pressure which only operate at pumping requirement pressures, there will be a greatly reduced incidence of refrigerant leaks. Calcium chloride brine is by far the most popular in use today. However there is controversy at the present concerning the corrosion inhibiter required for multi metal units. The existing chemical is alleged to be harmful to health and although there are alternatives proposed, the problem has, as yet, to be resolved. The major requirement for both the freezer and the refrigeration system on a fishing vessel is reliability.

Handling Fish Before Freezing

The layout of a stern trawler illustrates a good arrangement for handling fish before freezing. The fish are pulled up the stern ramp, poured from the net through hatches to the factory deck below and then moved forward as they go through the various stages of processing. Not all vessels can use this preferred arrangement and it would be impossible to cover all the potential layouts for the wide variety of vessels now used for freezing at sea.

The pre-freezing procedures described below are typical of a freezer trawler fishing in the North Atlantic. They may, however, have a more general application, and only minor modifications may be necessary to accommodate other vessels and their particular requirements. Fish must not be left lying on the upper deck exposed to direct sunlight but stored on a sheltered working deck immediately below. The fish should be kept as cool as possible immediately after catching and throughout the time they are awaiting freezing. Some means of cooling, such as a of seawater spray or chilled sea water (CSW) tanks, is therefore recommended when the fish are subject to any delay before gutting. Cooling of the fish not only helps to retard spoilage but also stops the blood from clotting too quickly. In tropical conditions, it will be necessary to provide a means of chilling the water used for this purpose.

Gutting of fish should commence as soon as possible after being caught not only to ensure the continuity of supply to the freezers, but also to reduce the rate of spoilage. The removal of the gut releases blood from the fish and must be followed by immediate washing in chilled water. Blood which is not released soon enough, clots within the tissues resulting in a permanent pink or red discoloration of the flesh which detracts from the appearance of the fillet. The liver must also be removed as it contains a fat which is highly perishable and could become rancid even at low temperatures. A time of 15 to 30 min in cold water is usually required for bleeding to be complete. However, in practice it is often difficult to ensure that all fish are given time to bleed properly. One solution is to make washing a two stage process. The gutted fish are first put into an open tank where they can bleed while they are kept cool by chilled water sprays, and then they are conveyed to an automatic fish washer where they are given a final rinse before freezing. The need for delay for proper bleeding of the fish before freezing must seem an added encumbrance. However, where it is desirable to gut before freezing, time must be allowed for the blood to be released from the fish to ensure they have good appearance. If appearance is not important in the final product, this delay for bleeding may not be necessary. Traditionally, some fish may only be marketable ungutted and in this case a special effort should be made to handle the fish quickly. They should be kept chilled and rough handling should be avoided.

Fish are usually sorted before freezing so that each block or package contains only one species. With some species, further subdivision into size grades may also be necessary. This extra handling for sorting and grading is viable when a premium is paid for graded fish. Some sorting is required before processing to reject unwanted and undersize fish and any trash material. Whether it is feasible at this time to sort the fish to be frozen into species and size grades will depend on individual requirements.

Heading of the fish may be desirable in some cases. Headed fish make a more compact block and heading allows larger fish to be frozen in some freezers. Removing the heads also means that the freezers can be used more efficiently and the proportion edible fish stored is also increased. One disadvantage of heading is that a small but not insignificant amount of edible fish is removed with the head. The cut surfaces may also become discoloured in time and trimming may be necessary.

Small pelagic fish such as herring are traditionally landed in the whole ungutted state and should also be kept cool since spoilage rate will be higher than for larger fish that have been gutted. Freezing if required should be done as soon as possible.

Another major problem, which usually only applies when fish are frozen at sea, is due to the effects of rigor mortis. Very often fish are bent prior to rigor. This should be avoided as far as possible since the flesh of the fish on the outside curve will be put under a strain and, when rigor sets in, the extra forces involved will pull the fish apart. This results in gaping of the fillets. The fillet in the inside of the curve, on the other hand, will shrink and contract so that two different looking fillets will be obtained form the same fish. One will be elongated with much gaping and the other will be short and compact. Freezing will of course maintain the fish in this bent condition, and will no doubt be blamed for the phenomenon. If bent fish in rigor are straightened before freezing, gaping of the short compact fillet will result. The onset of rigor mortis is quicker at higher temperatures and may occur only 10 to 20 min after death at temperatures near 30C. It is therefore essential that the fish be chilled quickly if problems due to rigor mortis are to be avoided during freezing.

If the fish are to be filleted at sea and frozen as fillets, it is even more important that chilled conditions exist along the entire production line. When a fish goes into rigor, there is a gradual increase in the tension of the muscle fibres and, as long as the muscle remains attached to the skeleton of the fish, shrinkage is restricted. However, once a fillet is cut from the fish, this restraint is removed and, if the onset of rigor mortis is not complete, the fillet will shrink. This contraction gives the fillet a corrugated appearance and a distorted shape. Temperature has an important bearing on this process, the higher the temperature, the faster the shrinkage and hence, the greater the effect in a given time. Fillets that are allowed to shrink at a high temperature before freezing can lose greater quantities of drip on thawing. In addition, excessive flexing or contact with water can increase the amount of shrinkage of a pre-rigor cut fillet.

A fillet taken from a fish before rigor has set in will, after freezing and thawing, have a dull appearance. This absence of gloss is probably due to the cut ends of muscle cells projecting upward. The fillet has a velvety feel and will not, for instance, produce an attractive smoked product. There is no known solution to this problem so far, apart from delaying filleting until after the onset of rigor mortis.

Before freezing, the fish may be stored in bins of an equivalent volume and adjacent to the freezers. This ensures that only the correct quantity of fish is available to fill the freezers and none are left on conveyors. The working space adjacent to the freezers should be kept cool to ensure that the fish do not warm up at this stage and the bins should be emptied in strict rotation so that no fish lie longer than necessary. Final sorting can be made at the freezer and a facility e provided for the storage or return of rejects.

Any chilling of the fish prior to freezing will not be an extra refrigeration requirement since any heat removed before entering the freezer means a reduction in the subsequent refrigeration loading. Chilling will be a low cost process with considerable benefits in improved fish quality.

Handling Frozen Fish

Fish should be transferred to the cold store immediately they are removed from the freezer. Even large blocks of fish warm up rapidly at ambient temperatures particularly in tropical climates. Heat added to the fish at this stage means that it has to be removed in the cold store and this will mean some loss in quality and an additional cold store load. Mechanical aids, such as chutes or conveyors, should be used since the less handling the frozen fish receives, the less chance there will be of the fish being damaged. Damaged blocks require more storage space and extra handling in the cold store. Damaged fish should be kept separate in the store since they may require special discharge arrangements.

Unfrozen or partially frozen fish should never be placed in the cold store as it is not designed for freezing. Partially frozen fish are also more easily damaged during handling. When fish are graded before freezing, the blocks or packages should be clearly labelled and, if possible, the different grades kept separate within the cold store. Labels placed on the surface of frozen fish and brushed over with clean water will adhere to the surface and this method of marking can be used if the fish are unwrapped. If wrappers are used for the fish, they should be suitable for marking so that the fish can be identified and handled more quickly at the time of discharge or later.

Cold Stores on Freezer Vessels

The cold store on a fishing vessel should operate at the temperature and on the same principles as recommended for shore based cold stores. Even if the storage time on the vessel is to be relatively short, it must be remembered that poor practice at any stage of handling and processing will have a cumulative effect which may become obvious by the time the fish reaches the consumer.

Loading and unloading of a fishing vessel's cold store is usually done through hatches at roof level. This is a good arrangement since there will be little exchange of air between the store and outside when the hatches are open. One distinct disadvantage of this hold arrangement is that even small refrigerant leaks can result in an accumulation of refrigerant in the cold store and, although the refrigerant may not be toxic, the resulting low oxygen level may be dangerous. Shipboard cold stores must therefore have an effective alarm system and the crew disciplined to use it and obey all other safety rules and regulations.

Insulation of a cold store on board a fishing vessel creates some special problems. The cold store insulation is usually attached directly to the ship's side; therefore, the rib structure of the vessel will penetrate into the insulation for some distance. Although the thickness of the insulation need not be increased to more than would be necessary for corresponding store on shore, the design should ensure that there is an effective thickness of insulation at all points in the store. Any internal structure within the cold store space should be attached to the main framework of the vessel with an effective heat barrier. Metal or any other material with a high thermal conductivity should not be used for this purpose.

Foamed-in-place polyurethane insulation and polyurethane slabs have been used for the insulation of shipboard cold stores. The application of this type of insulation is difficult and requires skilled operators and special equipment. Loose fill insulation may however be used in combination with others for instance for packing awkwardly-shaped areas where cutting of slabs would be difficult. Another desirable property of an insulation for a steel fishing vessel is that it should be reasonably heat-resistant to enable welding or other heat treatment to made to the outer hull. Unfortunately, none of the insulations that are likely to be used can completely satisfy all the requirements but some are significantly better than others. Insulation choice and method of application can only be made after consulting the relevant codes of practice or legislation for shipboard insulation for the country in which the vessel is registered or insured.

Increased insulation thickness can be costly on a fishing vessel since only a few centimetres increase in thickness can mean a significant reduction in the storage space. A store holding about 100t of fish, for instance, would be reduced to about 95t if the insulation thickness is increased by only 5cm.

The choice of cooling systems available for the cold stores of fishing vessels is much the same as that for other stores. Pipe grids and forced circulation coolers have both been used successfully. When large blocks of fish are stored, the system selected should not be vulnerable to damage due to block handling or movement.

Grids on the sides of the vessel require to be protected since a 45kg block of fish can damage pipework particularly since metal is a good deal more brittle at cold store temperature. Plain pipe grids on the roof only have been used but, in order to get the required heat transfer surface, finned grids are usually required and care must be taken to ensure that the frost on the pipes does not bridge the air space between the fins. If plain pipe grids are used, it will normally be necessary to continue the grids to at least halfway down the sides of the hold.

Attempts have been made to avoid using wall grids by having two or more rows of plain pipe grids on the roof. In terms of cold store quality, this is an inferior arrangement and also makes defrosting more difficult than when a single row only is used. Cooling grids together with any protective lining take up a good deal of space in a store but some of the space would normally be kept free as clearance around the produce.

As in some modern stores on shore, unit coolers may be used and they can be located outside the main storage area, for easier defrosting, and the cold air circulated uniformly within the room.

Unloading Freezer Vessels

Handling frozen fish is obviously different from handling iced fish and cannot be unloaded in the same manner. Any unloading system should handle the fish quickly between the vessel's cold store and the cold store on shore. Delays at this stage, particularly in warm climates, can result in partial thawing of the fish with a resultant loss in product quality.

There should be no delays on the quayside and any grading of the frozen fish according to species or size should be left until the produce is in the cold store or a least under cover. Ideally, cold stores should be adjacent to the landing place and fish can then be moved quickly, preferably by conveyor, into cold storage. Alternatively, if the store is reasonably near to the quay, unrefrigerated vehicles may be used for transporting the fish provided there are no delays. Vehicles however, should be of the enclosed type and they should be loaded under a canopy so that fish are not exposed to direct sunlight. Large capacity vehicles should not be used since extended loading times will result in a good deal of heat being added to the fish. Even when smaller vehicles are used, if delays are unavoidable the vehicle should be despatched to the coldstore with a partial load rather than wait to be fully loaded. In some countries, labour may be relatively cheap and mechanical handling may be considered an expensive luxury. In these cases, it may therefore be more economic and quicker to manhandle the fish. The rate of unloading of frozen fish from a vessel will depend on the size of blocks or packages or the containers used for loose fish. It will also depend on the facilities provided on the vessel, such as number and size of hatches and also the degree of accessibility and hence the number of men that can be employed at one time. With a mechanical unloading system and a skilled crew, unloading rates of 1 to 1.5 t/man hour can be achieved. Mobile cranes and ship's derricks have also been used to unload frozen fish into enclosed containers in the hold which are then transferred to the quayside. This operation should ensure that there is no delay on the quayside in order to transfer and sort fish. Storing the fish in containers in the ship's hold has been suggested but unless the vessel is specially designed for this purpose, up to 30 per cent of the available storage space could be lost due to the presence of the pallets and the need for squaring off the hold. This would mean that a fishhold capable of storing 600t of unpalleted frozen fish in open storage would only be able to hold 400t if the fish were graded and stored in containers. Only if a freezer vessel reached the proportions of a cargo ship would palletization be achieved without a significant loss since a vessel of that size would have parallel sides for a good length of the hold. The stowage rates given in Table 26 give some guidance for calculating the likely hold capacity.

Transporting of Frozen Fish

Frozen fish delivered to a destination where they are to be sold immediately are likely to be consumed within a few hours and no harm is done if they are partially thawed on arrival at their destination. The frozen fish may in fact be carried in uninsulated containers depending on how long the journey takes. Enclosed vehicles, however, should be used or at least a cover provided to protect the fish from direct sunlight. An insulated vehicle will be required for long journeys depending on the initial temperature of the fish, whether the vehicle is fully or partly loaded, the size of the load, the insulation quality and thickness, the degree of air ingress and the local climatic conditions. A local trial will ascertain the maximum range attainable.

Frozen fish that are to be transferred to other cold stores must be transported in an insulated vehicle preferably with some form of refrigeration equipment to maintain the air space at a temperature of approximately -20C. The following lists refrigeration methods that may be used:

1. Mechanical refrigeration using either wall coolers or forced convection coolers blowing air throughout the storage space. In some cases, a jacketed system for distributing the air is employed. This is the most common system.
2. Rechargeable eutectic plates.
3. Solid or liquid carbon dioxide or liquid nitrogen can be used with a total loss system.

The cost of a vehicle complete with a mechanical refrigeration system suitable for maintaining a temperature of -20C would be approximately US $110,000. This vehicle would be suitable for transporting 15t of frozen produce.

Prior to loading, the vehicle or container should be precooled and the loading should proceed quickly. Palletized loading and the formation of a scaled connection between the vehicle and cold store are both helpful in keeping the temperature rise at this stage to a minimum. The size of a package affects the speed at which it warms; the smaller the pack the greater is its surface area in relation to its volume and the quicker it warms. Fig 45 shows laboratory measurements made on a single consumer pack and on a carton of the same packs. Packaging the product in a master carton will clearly reduce the temperature rise during handling outside a refrigerated space.

Fish at the edges and corners of the load will warm more quickly than those at the centre of the load during unrefrigerated transport, and the extent of this temperature difference is not often appreciated by the operator. Fig 46 shows the result of temperature measurements made across the middle of a load in an uninsulated container. The temperature rise was almost entirely in the outer 300mm layer of the load which in this case was packed firmly against the container wall without an airspace. It must be remembered that the outer 300mm layer represents a considerable part of the total load. For example, in a container measuring 5 x 2 x 2m almost 60 percent of the load would be located within 300mm of the wall.

The above temperature measurements were made during the transport of frozen fish in a temperate climate where the ambient temperature was about 16C. The results clearly show the effects of bulk, size and position in a load on the rate of warming when no refrigeration is used. The differences will be even greater in warmer climates.

Refrigeration Plant Requirements

Most of the mechanical refrigeration plant used for freezing and cold storage of fish is of the vapour compression type basically consisting of compressor, condenser, expansion valve and evaporator (cooler). In simple terms, a refrigeration system takes in heat at a low temperature and rejects it at a higher temperature. The following is a brief description of the major components. Reference should be made to the appropriate text books for a more detailed description.

Compressors

The selection of a compressor to suit a particular installation is better left to a qualified person. Detailed information on compressor design cannot be given in a document such as this. As a general rule, small freezer and cold store installations should not share the same refrigeration machinery. Load fluctuations brought about by the freezer being loaded and unloaded, could result in temperature fluctuations within the cold store. In addition, when only the cold store only is in operation, a refrigeration compressor of large capacity will be used for what is a relatively small refrigeration load. Apart from being uneconomic, this will result in problems with capacity control. On the other hand large installations usually have multiple compressor systems and are maintained and controlled by competent engineers.

Condensers

Table 1 gives some indication of the water requirement for various types of condenser.

Table 1. Condenser water requirement (t/h)

Type of condenser	*100 kg/h freezer*	*1000 m^3 cold store*
Shell and tube (water rejected)	5 to 7	10 to 14
Shell and tube with water recooling	0.03 to 0.06	0. 06 to 0. 12
Evaporative	0.03 to 0.06	0.06 to 0. 12

Selection of a condenser must take into consideration many factors relating both to the system used and to the climatic conditions. Selection must again be left to a qualified person who is aware of all the relevant information about the project.

Duplication of Cold Store Plant

The value of frozen product in a cold store can be high and precautions must be taken to ensure that the contents are not damaged in the event of a major breakdown of the plant. Cooling by multiple units, each with a separate condensing unit, is one way of ensuring that sufficient refrigeration effect is available to maintain the store at the operating temperature or slightly higher should a unit break down. Another method is to cross-connect the cold store and freezer refrigeration pipework. This allows the freezer refrigeration machinery to be used to cool the store in an emergency. With normal operation, the two would be isolated and only a competent person would be allowed to make the cross-connection.

Centralised Plant

Centralised machinery enables one operator to take care of all the refrigeration equipment. Care should be taken in arranging plant layout such that the refrigeration lines to and from the cooler are not too long, as this can give rise to a number of difficulties. Plant operation and economics are considerations that also have to be taken into account.

Standardisation of Plant

Standardisation of equipment is another good policy to follow especially in remote areas. Parts can be interchangeable and stocks of spares will be kept low. If possible, the same refrigerant should be selected for each installation and similar machinery made by one manufacturer should be specified. The size of individual units should also be standardised whenever possible even if it means that some adjustment has to be made in their capacity to suit each requirement.

Simplicity and Reliability

Small plants seldom justify a full time engineer in attendance; therefore, simplicity and reliability should be major considerations when selecting the equipment particularly in a developing country. The plant and all auxiliaries should also be well tried and tested. Although these requirements apply particularly when plant is unattended, conditions in most developing countries are such that they should be applied there as a general rule. Whatever incentive there may be to purchase plant that is new and offers potential economic or other benefits, the purchaser should place a good deal of importance on reliability.

Power Requirements

The power required for the operation of a refrigerated warehouse is high-voltage main electricity. A transformer unit must supply low-voltage mains in the establishment via a general switchboard. The regulations in force in most countries imply that refrigerated premises act as good conductors due to moisture condensation risks and consequently impose a certain number of precautions. Those precautions include earthing of motors and all equipment through a special circuit connected to an earth-point that is independent of the high voltage earthing connector. Equipment should be waterproof with water-sealed cables, including those of lighting circuits. The supply for mobile equipment should be low voltage. The mains circuit within the store must include a conductor to earth the frames of fans, coolers, small machine tools and other equipment. An outdoor main contactor to switch off the entire installation except the engine room extractor fans should be provided for emergency purposes. With the increasing use of electrical trucks, provision for battery charging is required. Further plug-in facilities for refrigerated vehicles may be necessary. All work concerning the power supply must be carried out by specialists.

Refrigeration Plant Operators

Two people are important for operating a freezing plant and cold store; the engineer in charge of the refrigeration plant and equipment and the store operations manager or store keeper. Whatever the size, type or function of the plant a capable person should be employed to operate, maintain and repair all the equipment. The qualifications and ability of the operator required will depend on whether help is available locally to deal with major problems in the plant. Where skilled help is available locally or where the plant is small, a qualified refrigeration specialist is probably not required and recruits for this position need only be skilled or semi-skilled engineers who have experience in other industries. The engineer, however, has to be self-reliant, adaptable and able to make do with whatever facilities and materials are readily available to keep the plant operating. An ex-marine engineer for example, would be ideal particularly if he has had experience with steampowered plant. Only a minimum amount of training would be necessary to enable such a person to appreciate the particular problems that are applicable to refrigeration plant and he should also appreciate the reasons for good freezing and cold storage practice.

Even in industrial countries specialist training in refrigeration engineering is not widely available and it would be unreasonable to expect most developing countries to provide an organisation of their own for this purpose. However, qualified professional engineers have a broad-based training and where it is justified, attendance at a short course which deals with refrigeration and food technology may be all that is required. Fortunately, in most developing countries with a hot climate, there already exists a pool of technicians and plant operators who have gained refrigeration experience with ice-making and air conditioning plant. Recruitment from this source would mean that only the minimum amount of training would be necessary to enable them to operate new plant.

The recruitment of a qualified store operations manager should not be difficult. This person should be able to keep records of the movements of goods in and out of the store, be responsible for stowage of the goods and be able to keep simple accounts and deal with dispatch and invoice notes. An efficient store manager could considerably reduce the handling costs and improve the utilisation of the plant and storage space. The person selected for this post should therefore have good organisational ability and have experience in a similar post, but not necessarily connected with the refrigeration industry. Additional training for this person, should give him an appreciation of the perishable nature of the goods he is handling and a knowledge of appropriate cold store practices for maintaining the quality of frozen fish products.

References

Aitken,A *et al.* (eds) 1982. *Fish Handling and Processing*. Second Edition, Edinburgh, Her Majesty's Stationery Office, £10.

FAO, 1975. Ice in Fisheries. *FAO Fish Rep.* (59) Rev. 1 (issued also in French and Spanish), 57p.

International Institute of Refrigeration. *Guide to Refrigerated Storage.* Guide de Paris,

Jenkins, C.H. 1968. *Modern Warehouse Management.* New York, McGraw-Hill Book Company

Myers, M., 1981, Planning and Engineering Data. 2. *Fresh Fish Handling*. FAO Fish. Circ., (735), 64p.

5

Food Fermentation Technology

Food microbiology can bedivided into three focus areas; beneficial microorganisms, spoilage microorganisms, and disease causing microorganisms. Beneficial microorganisms are those used in food fermentation to produce products such as cheese, fermented meat (pepperoni), fermented vegetables (pickles), fermented dairy products (yoghurt), and ethnic fermented products such as sauerkraut, idli and kimchi.Infermented products (producedbynaturalorcontrolled fermentation), microorganisms metabolize complex substrates to produce enzymes, flavor compounds, acids, and antimicrobial agents to improve product shelf-life and to prevent growth of pathogens and to provide product attributes. Micro organisms with their enzymes, also breakdown indigestible compounds to make the product more palatable and easy to digest. In addition, the beneficial microorganisms also serve as probiotics to impart direct health benefit by modulating the immune system to provide protection against chronic metabolic diseases, bacterial infection, atherosclerosis, and allergic responses. Examples of beneficial microorganisms are *Lactobacillus acidophilus, Lactobacillus arabinosus, Lactobacillus lactis,* and *Pediococcus cerevisiae.* Food spoilage microorganisms are those which upon growth in a food, produce undesirable flavour (odour), texture and appearance, and make food unsuitable for human consumption. Sometimes uncontrolled growth of many of the beneficial microorganisms can also cause spoilage. Food spoilage is a serious issue in developing countries because of inadequate processing and refrigeration facilities. Examples of food spoilage microorganisms are *Brochotrix, Lactobacillus, Bacillus*, *Pseudomonas* spp. and some molds. The micro-environment created in a spoilt food generally discourages the growth of the pathogenic microorganisms, which are considered poor competitors.

FERMENTATION

The uniqueness of several microorganisms and their often unpredictable nature and biosynthetic capabilities, in a specific set of environmental conditions, have made them ideal candidates, in attempts to solve difficult problems in life sciences and other fields. Microorganisms have been used in various ways over the past many decades, to advance medical technology, human and animal health, food processing, food safety and quality, genetic engineering, environmental protection and agricultural biotechnology. The use of beneficial microorganisms in the food sector has been a long tradition, namely lactic acid bacteria and yeasts in fermentation processes; the former are widely usedinthe manufactureoffermented food and are among the best studied microorganisms. The fermentations may be byyeasts, bacteria, molds or combination of these organisms. Detailed knowledge of a number of physiological traits has opened novel potential applications for these organismsinthe food industry, while other traits might bespecifically, beneficial for human health.

The food and beverage industry exploits non-pathogenic microorganisms for the production of fermented foods. These foods are prepared from raw materials and acquire their characteristic properties by a process that involves microorganisms. In certain cases the endogenous enzymes of the wild micro flora nature to the raw material may play a decisive role. It is believed that fermented foods originated from the Orient and date back to the prehistoric times. Originally, these were fermented *"spontaneously"* by autochthonous strains found in the raw materials or the environment; This was the start of traditional biotechnology. The most important were cheese, yoghurt, wine, vinegar, beer, bread and the traditional fungal fermentations used inAsia andAfrica, for the production of food.

In the biochemical sense, the term fermentation refers to a metabolic process in which organic compounds (particularly carbohydrates) are broken down to release energy without the involvement of a terminal electron acceptor such as oxygen. Partial oxidation of the substrate occurs so that only a relatively small amount of ATP energy is released compared to the energy generated if a terminal electron acceptor is involved. Partial oxidation of a carbohydrate can give rise to a variety of organic compounds suchasalcohols organic acids and acetone. The compounds produced bymicro-organisms vary from organism to organism and are produced through different metabolic pathways.

The term fermentation can also be applied to any industrial process that produces a material that is useful to humans and if the process depends on the activity of one or more micro-organisms (consortia). These processes, known as industrial fermentations, are usually carried out on a large scale and in vessels in which the organisms are normally grown in liquid media. Some industrial fermentations are fermentation inthe biochemical sense but the majorityof microorganisms involved are aerobic and use oxygen (as terminal electron acceptor) and thereby metabolize carbohydrates completely.

A vast range of materials are produced by industrial fermentations. These include

1) Organic chemicals used as fuels, food additives, antibiotics and enzymes for use in the food and other industries. Vinegar is an example of a food additive produced by an industrial fermentation.
2) Organisms are produced on a large scale for the extraction of protein that can be used as a part of the human diet. Quorn is an example of a single cell protein, It is produced from the fungus *Fusarium graminearum*. This mycoprotein, purified from the fungus,iscurrently available for use as a food and is incorporated into a range of dishes that appear on supermarket shelves. Meatless dishes with a high protein content made from Quorn are particularly appealing to vegetarians.
3) Yeast cells produced for use in industries such as the baking industry, which relies on the mass production of large amounts of baker's yeast.
4) Foods produced on a large scale as a result of the activities of micro-organism, e.g. cheese, yogurt and bread.
5) Production of alcoholic beverages, e.g. beer and wines.
6) Cellular extracts used as food additives, e.g., yeast extracts from yeast cells produced as a by product of the brewing industry.
7) Industrial fermentations are now often considered under the heading of biotechnology, i.e. technology that uses living organisms and their products in the manufacturing and service industries.

Fermented Foods and their Importance

Fermented foods are those foods produced by the modification of a raw material of either animal or vegetable origin by the activities of micro-organisms. Bacteria, yeast and moulds can be used to produce a diverse range of products, that differ in flavour, texture and stability from the original raw material. The production of many fermented foods involves organisms that are biochemically fermentative. Lactic acid bacteria that ferment carbohydrates to produce lactic acid are particularly important, but yeasts also play a major role in some food fermentations, fermenting carbohydrates to produce ethanol and other organic chemicals. Moulds that do not ferment carbohydrates, also play an essential role in some food fermentations, for example, the production of cheeses (blue cheeses) and oriental foods (soy sauce). Fermented foods are an extremely valuable addition to the human diet for a whole variety of reasons:

1) *Increase in variety:* Fermented foods increase the variety of foods that are available, adding to our diet a group of highly nutritious products with unique characteristics. There are, for example, about 1000 different types of cheeses.

2) *Use of ingredients:* Fermented foods form an important ingredient for a wide variety of dishes and are often used to impart special flavours, e.g. pepperoni in pizzas, yoghurt in curries, cheeses in a whole range of dishes, including soups, and soy sauce in stir-fry dishes.

3) *Improvement in nutritional quality:* The fermentation process may improve the nutritional quality of a raw material. Here are some examples:

    — Tempeh fermentation raises the vitamin B 12 content of the original soybean.

    — Tapioca fermentation doubles the protein content of cassava and increases the level of essential amino acids.

    — The presence of yeasts in a fermented food will increase the vitamin B content.

    — Antinutritional factors such as phytates, glucosinolates and lectins may be removed by the fermentation process.

    — Fermentation may lead to an increase in the bio-availability of minerals.

These improvements in the nutritional value of raw material will have little effect in the balanced diets of Western populations. However, for populations that subsist on diets consisting largely of polished rice, maize or other starches, such as in Africa and Asia, the contribution that fermented foods make to the intake of B group vitamins and proteins is highly significant.

4) *Preservation*: Fermentation often preserves a raw material, improving safety with regard to food-borne pathogens and increasing shelf-life; compare the shelf-life of raw milk (only a few days) with the shelf-life of yoghurt (several weeks).

5) *Health benefits:* Some fermented foods are said to have definite health benefits, although the scientific evidence for this is limited. Reports suggest that fermented milk products such as yoghurt can reduce serum cholesterol levels and help avoid cancers, particularly, those associated with the colon. 'Bio'yoghurts (AB and ABT yoghurts) are said to have a restorative effect on a normal micro flora, assisting recovery of a normal balanced flora after oral antibiotic therapy.

6) *Improved digestibility*: Some fermented foods are more easily digested than the original raw material. People who cannot digest lactose properly (show lactose intolerance) can often consume some types of fermented dairy products (particularly yoghurts) without harmful effects. Lactose intolerance is due to the absence of the enzyme galactosidase in digestive juices, which converts lactose to glucose and galactose. Ingestion of dairy products leaves unabsorbed lactose in the gut, which is fermented by the normal gut flora giving flatulence, abdominal pain and diarrhoea. The fermentation of milk converts the difficult to digest lactose to the more easily digested lactate, and the galactosidase in live starter culture organisms appears to assist in the digestion of any residual lactose. Legumes, e.g. soybean, contain oligosaccharides such as stachyose which are fermented in the gut to yield gas and the associated socially embarrassing flatus. The oligosaccharides are broken down to readily digestible monosaccharides and disacccharides during fermentation of legumes bymoulds, thus removing the problem.

7) *Detoxification of raw materials :* The fermentation process may remove toxic chemicals presentinthe raw material. Cassava fermentation, for example, removes a cyanogenic glycoside; cassava is toxic if eaten raw.

Fermentation Science and Technology

The science of fermentation is known as zymology. Food fermentation involves all those fermentation processes where either the ultimate product is used directly as a food, as an additive to food or is a basic ingredient to the food or the by-product formed during fermentation, food waste utilization, their disposal or proper management. As a science the food fermentation, has an element of biological sciences especially the microbiology, genetics and biochemistry, as a technology the food technology, chemical engineering along with integral component of sciences involved in food toxicity, acceptability and food nutrition. For any process to make fermented food, essential components are raw materials, micro-organism(starter cultures), fermentation vessel and associated controls, processing, recovery and packing systems, but from fermentation technology point of view, the fermenter and micro-organism involved assumes prime importance. Since the end product would serve as a food, it is essential to evaluate it from nutritional toxicology and sensory quality point of view and thus, the knowledge of these aspects would form an integral part of food fermentation technology, The technological dimensions of food fermentation towards the application of engineering sciences in designing the fermenters and associated controls for optimum fermentation and product recovery are immense. As the product on the commercial scale would be marketed, which can occur only if it is economical and therefore the economics of such products and the associated marketing aspects cannot be ignored. Similarly, transfer of the knowledge to the scale of technology operation required is of paramount importance. The scale up operations, harvesting, biocatalysis of micro-organisms, product recovery, effluent treatment etc. are of concern to fermentation technologists. Food fermentation technology could be organized based either on the product or the system used to make the fermented foods or products derived from it similar to food biotechnology.

Types of Food Fermentations

A number of different types of food fermentation can be recognized.

Acid Food Fermentation

These include acid fermented dairy products, e.g. cheese, butter, yoghurt and kefir; acid fermented vegetable products, e.g. sauerkraut, olives and various pickles; acid fermented meat products, e.g. the semi-dry fermented meats such as cerevelat and the dry fermented meats such as salami and pepperoni; sour dough breads.

The common feature of all these products is the use of lactic acid bacteria to carry out the basic fermentation process. Modern production systems usually involves the use of starter cultures.An exception is the fermentation of sauerkraut for which the process depends on lactic acid bacteria, that are natural inhabitants of the surface of cabbage leaves. Sometimes sugar is added to raw material to allow the lactic acid bacteriatoproduce sufficient acid forasuccessful fermentation. This is the case with fermented meats in which the sugar content of the raw material is very low. Salt may be added to suppress the growth of the normal spoilage micro flora and allow the lactic acid bacteria to dominate, e.g. sauerkraut, pickles and fermented meats. The raw materials may be pasteurized to eliminate pathogens and suppress natural contaminants that compete with the lactic acid bacteria, used in the starter culture.

Yeast Fermentation

Yeasts are important in food fermentation because of their ability to produce carbon dioxide and ethanol. Carbon dioxide is the important metabolic product in the manufacture of leavened bread whereas ethanol is metabolized in the production of beer, wine and spirits.

Solid State Fermentation

Solid state fermentation involves the use of a solid substrate into which the fermenting organism is inoculated. The organisms used are often molds. Examples are the 'koji' process and the second stage of tempeh fermentation.

Common Examples of Food Fermentations

Avery wide range of inumerable products of the food industry, such as sour cream, yoghurt, cheese, fermented meat, bread and other bakery products, alcoholic beverages, vinegar, fermented vegetables and pickles, etc., are produced through microbial fermentation processes. The efficiency of the strains of the organisms used, and the processes are being continuously improved to market quality products at more reasonable costs.

Oriental and Indigenous Fermented Foods

A large number of fermented foods can be grouped under the heading of oriental and indigenous fermented foods. Fermented foods of this type are produced in Asia and Africa and are often associated with specific countries or areas. Most of the products are unknown in the West but frequently have maj or nutritional role in the diets of the local population. Lactic acid bacteria are involved in some of the fermentation but yeast and moulds are often the main organisms responsible. Many are solid state fermentations or involve fermentations of more than one type. Some of the products are manufactured on large scale but many are carried out on a cottage industry or household basis.

Fermented Vegetable Foods

a) Sauerkraut

Sauerkraut is fermented fresh cabbage product. It is popular in USSR and Europe. The main organism involvedinthe fermentationofthis pickleis lactic acid bacteria, *Leuconostoc mesenteroides* followed by *Lactobacillus plantarum*.

b) Cucumber pickle

Cucumber pickle is a fermentation product of fresh cucumbers. Several lactic acid bacteria are involved in preparation of this pickle. *Lactobacillus plantarum* is the most important organism required for fermentation of cucumber pickle.

Fermented Soyabean Products

a) Tempeh

Tempeh is a highlypopular soyabean preparation in Indonesia. The chief organism in this preparation is the mold *Rhizopus oligosporus*. The boiled soybean seeds are mashed and wrapped in banana leaves or kept in boxes or hollow tubes. It is inoculated with spores of tempeh fungus by addition of a portion of previous batch and allowed to ferment for about 20 hours at a temperature 32°C until there is a good growing mycelium but little sporulation. It is then sliced and prepared as per the taste such as roasting or frying. The taste of the tempeh is considered to be bland but it is highly nutritious.

b) Soya sauce

Soya sauce is a very popular preparation of Japan, which has received wide acceptance world over. This is prepared by inoculating *Aspergillus oryzae*

(A. Soyae) in a mixture of soaked and steamed soy bean with roasted wheat in the ratio of 2:1. The mixture is incubated at 25-30^{0}C for a period of 3 to 5 days. Subsequently, it is subjected to various processing steps using bacterium *Lactobacillus delbruckii* and the yeast *Saccharomyces rouxii.* After 3 months, the final product is filtered, pasteurized and bottled for use.

c) Miso

It is made from fermented soyabeans and is a thick paste-like substance. Miso is brownish in color and tastes extremely saltyand tangy on its own. Saccharomyces rouxii and Torulopsis are the yeasts and Pediococcus halophilus and Streptococcus faecalis are the bacteria, principally involved in Miso fermentation. While the most common use of miso is in Japanese-style miso soup recipes, miso also adds a unique burst of flavour to salad dressings, sauces and marinades, baked tofu, or vegetable dishes.

Fermented Dairy Products

The fermented dairy products assume greater importance in the human diets as invariably most of the diet consist of milk products especially the cheese, butter, yoghurt, curd etc. There are a number of fermented dairy products as shown below:

Economically Important Fermentation Products

The growth of micro-organisms or other cells results in a wide range of products. Each culture operation has one or few set objectives. The process has to be monitored carefully and continuously, to maintain the precise conditions needed and recover optimum levels of products. Accordingly, fermentation processes aim at one or more of the following:

a) production of cells (biomass) such as yeasts;

b) extraction of metabolic products such amino acids, proteins (including enzymes), vitamins, alcohol, etc., for human and/or animal consumption or industrial use such as fertilizer production;

c) modification of compounds (through themediation of elicitors or through bio transformation); and

d) production of recombinant products

Microbial Biomass

Microbial biomass is produced commercially as single cell protein (SCP) using such unicellular algae as species of Chlorella or Spirulina for human or animal consumption, or viable yeast cells needed for the baking industry, which was also used as human feed at one time. Bacterial biomass is used as animal feed. The biomass of *Fusarium graminearum* is also produced for a similar use.

Microbial Metabolites

Primary metabolites

During the log or exponential phase organisms produce a variety of substances that are essential for their growth, such as nucleotides, nucleic acids, amino acids, proteins, carbohydrates, lipids, etc., or by-products of energy yielding metabolism such as ethanol, acetone, butanol, etc. This phase is described as the trop-phase, and the products are usually called primary metabolites. Commercial examples of such products are given in Table 1.

Table 1: Examples of Commercially Produced Primary Metabolites

Primary Metabolite	*Organism*	*Significance*
Ethanol	*Saccharomycescerevisiae*	alcoholic beverages
	Kluyzveromycesfragilis	
Citric acid	*Aspergillus niger*	food industry
Acetone and	*Clostridium*	
butanol	*acetobutyricum*	solvents
Lysine	*Corynebacterium*	nutritional additive
Glutamic acid	*glutamacium*	flavour enhancer
Riboflavin	*Ashbyagossipii*	nutritional
	Eremothecium ashbyi	
Vitamin B12	*Pseudomonas denitrificans*	nutritional
	Propionibacterium shermanii	
Dextran	*Leuconostoc mesenteroides*	industrial
Xanthangum	*Xanthomonascampestris*	industrial

Secondary metabolites

Organisms produce a number of products, in addition to the primary metabolites. The microbial growth phase, during which products that have no obvious role in metabolism of the synthesizing culture organisms are produced, is called the idiophase, and these products are called secondary metabolites. In reality, the distinction between the primary and secondary metabolites is not a straight jacket situation. Many secondary metabolites are produced from intermediates and end products of secondary metabolism. Some like the *Enterobacteriaceae* do not undergo secondary metabolism. Examples of secondary metabolites are given in Table 2.

Table 2: Examples of Commercially Produced Secondary Metabolites

Metabolite	*Species*	*Significance*
Penicillin	*Penicilliumchrysogenum*	antibiotic
Erythromycin	*Streptomyces erythreus*	antibiotic
Streptomycin	*Streptomyces griseus*	antibiotic
Cephalosporin	*Cephalosporiumacrimonium*	antibiotic
Griseofulvin	*Penicillium griseofulvin*	antifungal antibiotic
CyclosporinA	*Tolypocladium inflatum*	immunosuppressant

Production of Enzymes

Industrial production of enzymes is needed for the commercial production of food and beverages. Enzymes are also used in clinical or industrial analysis and now they are even added to washing powders (cellulase, protease, lipase). Enzymes may be produced by microbial, plant or animal cultures. Even plant and animal enzymes can be produced by microbial fermentation through techniques of genetic manipulations. While most enzymes are produced in the tropophase, some like the amylases (by *Bacillus stearothermophilus*) are produced in the idiophase, and hence are secondary metabolites. Examples of enzymes produced through fermentation processes are given in Table 3.

Table 3. Examples of Commercially Produced Enzymes

Organism	*Enzyme*
Aspergillus oryzae	Amylases
Aspergillus niger	Glucoamylase
Trichodermareesii	Cellulase
Saccharomyces cerevisiae	Invertase
Kluyzveromycesfragilis	Lactase
Saccharomycopsis lipolytica	Lipase
Aspergillus species	Pectinases and proteases
Bacillus species	Proteases
Mucorpusillus	Microbial rennet
Mucormeihei	Microbial rennet

Fermented Foods as Functional Foods

Dr. Elias Metchnikoff, a Nobel Laureate, around 100 years ago proposed that *"Lactic acid bacteria can render a great service in the fight against intestinal putrefaction"* and might *"postpone and ameliorate old age"*. This concept was developed further through the decades, and today, the emerging probiotics, prebiotics and synbiotics era is a subject of scientific debate and intense research. The use of probiotics, prebiotics, and synbiotics is a promising area for the development of functional foods.

Functional foods are generally characterized as foods similar in appearance to conventional foods, consumed as part of a usual diet, and providing health-related benefits beyond meeting basic nutritional needs. A food can be considered naturally 'functional' if it contains a food component that affects one or more targeted functions in a beneficial way. Dairy foods can be included in the functional food category because of their content of calcium, specific health-enhancing proteins, conjugated linoleic acid, sphingolipids, butyric acid, and probiotic cultures.

Dairy foods appear to be the preferred medium for introducing probiotic bacteria such as human-derived species of lactic acid bacteria (e.g., *L. acidophilus, L. casei, L. gasseri, L. rhamnosus, L. reuteri, Bifidobacterium bifidum, B breve, B. infantis* and *B longum). Lactobacillus spp.* (naturally found in the human small intestine) and various *Bifidobacterium spp* (a major organism in the human large intestine) are the most commonly used probiotic cultures.

Probiotics

The term Probiotic means "for life", is derived from the Greek language. It is now defined as "live microorganisms which when administered in adequate amounts confer a health benefit on the host". The most common probiotic organisms clinically useful include members of the bacteria such as *Lactobacillus and Bifidobacteria* species and some selected strains of *Streptococcus, Lactococcus* and members of yeast and molds such as *Saccharomyces, Aspergillus, Acanthosis* and *Candida* species.

Use of Probiotics

As per definition of probiotics the live microorganisms that are administered have to be in adequate amounts to confer a health benefit to the host. Clinical bodies have used probiotic levels of 1 billion to 10 billion or above.

Dosage

According to Earl Mindell, an internationally recognized expert on nutrition, healthy person should take 2-5 billion CFUs (colony stimulating units) of probiotics a day and those with GI conditions can take up to 10 billion CFUs per day. In acute infectious diarrhoea, lactobacillus is most effective at a dose of 10 billion CFUs during the first 48 hours, which translates to 5 billion CFUs per day. For prescription probiotics, the current daily intake recommended is 5-10 billion CFUs per day. Capsules and Sachets of Probiotic plus Prebiotic combination (Pro-wel) and Probiotic alone (Darolac) are commercially available. Benefits offered by Probiotic and Prebiotic combination formula are:

— *Maximum Colony Forming Units (CFUs):* ensure complete action.

— *Fructo-oligosaccharide (FOS):* offers nutrition to the probiotics and normal intestinal flora.

— *Acid-resistant cells*: reach intestine in full force.

— *Freezed-dried and nitrogen-flushed cells:* offers excellent stability.

— *Vegetable capsules:* ensure universal appeal.

Health Benefits of Probiotics

Few well designed, well conducted human clinical trials of probiotics have been conducted over the past 30 years. Only in recent years, has the importance of choosing probiotic strains of demonstrated efficacy been recognized. With respect to health benefits of probiotics, research studies indicate the following:

1) Improved intestinal health.
2) Modulationof the immune response and immune function.
3) Reduced risk of cancer.
4) Reduced risk of heart disease.
5) Improved tolerance to milk.
6) Reduce food allergies.
7) Reduce ulcers by decreasing the growth of ulcer-inducing bacteria (i.e., *Helicobacter pylori).*

Fermentation is a relatively efficient, low energy preservation process which increases the shelf life and decreases the need for refrigeration or other forms of food preservation technologies. It is therefore a highly appropriate technique for use in developing countries and remote areas where access to sophisticated equipment is limited. Fermented foods are popular throughout the world and in some regions, make a significant contribution to the diet of millions of individuals.

In Asia the preparation of fermented foods is a widespread tradition. The fermented products supply protein, minerals and other nutrients that add variety and nutritional fortification to otherwise starchy and bland diets. For instance Soy sauce is consumed throughout the world and is a fundamental ingredient in diets from Indonesia to Japan. Over one billion litres are produced each year in Japan alone. Although fermentation of foods has been in use for thousands of year, it is likely that the underlying microbial and enzymatic processes responsible for the transformations were largely unknown. It is only recently that there has been a development in the understanding of these processes and their adaptation for commercialization. There is tremendous scope and potential for the use of micro-organisms towards meeting the growing world demand for food, through efficient utilization of available natural food and feed stocks and the transformation of waste materials. Genetic improvementofthe organismisfundamentaltothe success of fermentation technology. Mutation and recombination are the two ways to meet this end.

The alarming increase in inappropriate use of antibiotics and enhanced bacterial resistance, along with renewed interest in ecological methods to prevent infections makes probiotics, prebiotics and synbiotics a very interesting field for research.

Evidence is beginningto accumulate, describing their beneficial effects in avariety of GI (Gastro Intestinal) and non-GI disorders. They offer dietary means to support the balance of the intestinal flora. They may be used to counteract local immunological dysfunction, to stabilize the intestinal barrier function, to prevent infectious succession of pathogenic microorganisms and to influence intestinal metabolism. Looking ahead, this field holds immense promise for the future in delivering novel therapiesin different fields.

Prebiotics

The term Prebiotic is defined as "short chained carbohydrates that are indigestible by human enzymes in the GIT (Gastro-intestinal tract) and sclectively stimulate the growth and activity of specific species of bacteria in the gut, usually bifidobacteria and lactobacilli, with benefits to health".

The most commonly used prebiotics in supplements are Fructo-oligosaccharides (FOS). Bifidobacteria, due to the presence of beta-fructofur-anosidase enzyme are liable to break down and utilize

FOS. This helps in stimulation of bifidobacterium growth in the GIT. FOS exhibits nutritional properties on colonic pH and stool bulking. It also increases bioavailability of essential minerals and decreases serum triglycerides.

The other types of prebiotic substrates include-

— Xylitol, Sorbitol, Mannitol
— Disaccharides-Lactulose,Lactilol
— Oligosaccharides-Raffinose, Soybean, Palatinose, Isomaltose, Lactosucrose
— Polysaccharides-In-ulin, resistant starch

Synbiotics

The term synbiotic is used when a product contains both probiotics and prebiotics. Since the word alludes to synergism, this term should be reserved for products in which the prebiotic compound selectively favors the probiotic compound, e.g., FOS in combination with strains such as Bifidobacterium *B infantis, B longum* etc. Combining probiotics with prebiotics could improve the survival of the bacteria crossing the upper part of the GIT, thus enhancing their effects in the large bowel. Moreover, the local and the systemic beneficial effects of probiotics and prebiotics might be additive or even synergistic.

References

Banwart, G.J. 1979. *Basic Food Microbiology*, AVI Publishing Co. Inc., Westport, Connecticut.

Frazier, W.C. and Westoff, D.C. 1996. *Food Microbiology,* Tata McGraw Hill Publishing Co. Ltd., New Delhi.

Pelczar, M. Jr., Chan, E.C.S. and Kreig, N.R. 1993. *Microbiology*, Tata McGraw Hill Inc., New York.

Garbutt, J. 1998. *Essentials of Food Microbiology*, Arnold International Student's Edition, London.

6

Pickling

Pickling, also known as brining or corning, is the process of preserving food by anaerobic fermentation in brine to produce lactic acid, or marinating and storing it in an acid solution, usually vinegar (acetic acid). The resulting food is called a pickle. This procedure gives the food a salty or sour taste. In South Asia, edible oils are used as the pickling medium with vinegar

Another distinguishing characteristic is a pH less than 4.6, which is sufficient to kill most bacteria. Pickling can preserve perishable foods for months. Antimicrobial herbs and spices, such as mustard seed, garlic, cinnamon or cloves, are often added. If the food contains sufficient moisture, a pickling brine may be produced simply by adding dry salt. For example, sauerkraut and Korean kimchi are produced by salting the vegetables to draw out excess water. Natural fermentation at room temperature, by lactic acid bacteria, produces the required acidity. Other pickles are made by placing vegetables in vinegar. Unlike the canning process, pickling (which includes fermentation) does not require that the food be completely sterile before it is sealed. The acidity or salinity of the solution, the temperature of fermentation, and the exclusion of oxygen determine which microorganisms dominate, and determine the flavor of the end product.

When both salt concentration and temperature are low, Leuconostoc mesenteroides dominates, producing a mix of acids, alcohol, and aroma compounds. At higher temperatures Lactobacillus plantarum dominates, which produces primarily lactic acid. Many pickles start with Leuconostoc, and change to Lactobacillus with higher acidity.

Pickling began 4000 years ago using cucumbers native to India. It is called "achar" in northern India. This was used as a way to preserve food for out-of-season use and for long journeys, especially by sea. Salt pork and salt beef were common staples for sailors before the days of steam engines. Although the process was invented to preserve foods, pickles are also made and eaten because people enjoy the resulting flavors. Pickling may also improve the nutritional value of food by introducing B vitamins produced by bacteria.

Pickling Process

In chemical pickling, the jar and lid are first boiled in order to sterilize them. The fruits or vegetables to be pickled are then added to the jar along with either brine or vinegar or both, as well as spices, and are then allowed to ferment until the desired taste is obtained.

The food can be pre-soaked in brine before transfering to vinegar. This reduces the water content of the food which would otherwise dilute the vinegar. This method is particularly useful for fruit and vegitables with a high natural water content.

Dry Salted Pickles

Dry salting is used for pickling many vegetables and fruits including limes, lemons and cucumbers. For dry salt pickling any variety of common salt is suitable as long as it is pure. Impurities or additives can cause problems:

- Chemicals to reduce caking should not be used as they make the brine cloudy.
- Lime impurities can reduce the acidity of the final product and reduce the shelf life of the product.
- Iron impurities can result in the blackening of the vegetables.
- Magnesium impurities impart a bitter taste.
- Carbonates can result in pickles with a soft texture.

Dry Salted Lime Pickle

Location of production

Dry salted lime pickles are produced in Asia and Africa. They are particularly popular in India, Pakistan and North Africa.

Product Description

With dry salting, the limes are treated with dry salt. The salt extracts and juice from the vegetable and create the brine. The final product is a sour lime pickle. Spices are added depending on local preference. In India and Pakistan, the pickle is usually very spicy and hot due to the addition of chilli. It is usually eaten as a condiment.

Preparation of raw materials

The limes need to be selected and prepared. Only fully ripe limes without bruising or damage should be used. All limes need to be washed in potable cold water, drained, and then cut into quarters. Spices should be of good quality and free of mould.

Processing

Limes are placed in a layer, approximately 2.5 cm deep, into the fermenting container (a barrel or keg). One kilogram of salt is added for every four kilograms of limes. The salt is sprinkled over the vegetables. Another layer of vegetables is added and more salt added. This is repeated until the container is three quarters full. A cloth is placed above the vegetables and a weight added to compress the vegetables and assist in the formation of a brine. The formation of a brine takes about 24 hours.

As soon as the brine is formed, fermentation starts. As fermentation starts bubbles of carbon dioxide appear. Fermentation takes between one and four weeks depending on the ambient temperature. Fermentation is complete when no more bubbles appear.

Flow diagram

Selection. Only ripe limes should be selected
↓
Wash. In clean water
↓
Cut. Cut into four pieces or slice the skin
↓
Mix with salt. 1kg salt for 4kg of limes
↓
Ferment. Leave the container in the sun for a week to ferment
↓
Package

Packaging and storage

The vegetables can be removed from the brine and packaged in a variety of mixtures which may consist of vinegar and spices or oil and spices. Lime pickle can be packed in small polythene bags and sealed or in clean jars and capped. Lime pickles keep well if stored in a cool place. Due to the high acid level of the final product, the risk of food poisoning is low.

Pickled Cucumbers

Location of production

Pickled cucumbers are made in Africa, Asia and Latin America.

Product description

Cucumbers undergo a typical lactic acid fermentation and change from a pale product to a darker green and more transparent product.*Khalpi* is a cucumber pickle popular during the summer months in Nepal.

Raw material preparation

Fully ripe cucumbers without bruising or damage are washed in potable cold water and drained. The cucumbers can be pickled whole or sliced. With *khalpi* the cucumbers are washed, sliced and cut into 5-8 cm pieces.

Processing

1 kg of salt is added to every 20 kg of small cucumbers and 15 kg of large cucumbers. The brine should be formed within 24 hours by osmosis. If the brine formed by osmosis does not cover the cucumbers 40° Salometer brine is added to the desired level. A day or two after the tank is filled and closed the brine should be stirred in order to help equalise the concentration of salt throughout the mass.

As soon as the brine is formed, fermentation starts and bubbles of carbon dioxide appear. Fermentation takes between one and four weeks depending on the ambient temperature. Fermentation is complete when no more bubbles appear.

During fermentation the brine becomes cloudy for the first few days due to the growth of bacteria. Later if the brine is not covered, a filmy yeast growth will often occur on the surface.

Flow diagram

Selection. Only ripe cucumbers should be selected

↓

Wash. In clean water

↓

Mix with salt. 1kg salt for 15-20kg of cucumbers

↓

Ferment. For between one and four weeks

↓

Package

Packaging and storage

Cucumber pickle is usually stored in clean capped jars. They keep well if stored in a cool place. Due to the high acid level of the final product, the risk of food poisoning is low. With *khalpi* in Nepal, oil is added.

Pak-Gard-Dong (Pickled leafy vegetable)

Pak-Gard-Dong is a fermented mustard leaf (*Brassica juncea*) product made in Thailand. The mustard leaves are washed, wilted in the sun, mixed with salt, packed into containers for 12 hours. The water is then drained and a 3% sugar solution added. They are again allowed to ferment for three to five days at room temperature. Micro-organisms associated with the fermentation include *Lactobacillus brevis, Pediococcus cerevisiae* and *Lactobacillus plantarum.*

A similar product (*Hum choy*) is made in the South of China. This is produced by fermenting a local leafy vegetable. The leaves are washed and drained. They are then covered in salt and hung on racks to dry in the sun. The wilted leaves are placed in earthenware pots and covered with rice water, obtained after washing rice grains. The pots are sealed and the leaves allowed to ferment for four days. The product can be stored for up to two months if the seal is not broken.

Tempoyak (pickled durian)

Tempoyak is the fermented pulp of a durian fruit (*Durio zibethinus*) from Malaysia. It has the distinctive durian smell and a creamy yellow colour. It is made by mixing durian pulp with salt and placing in a sealed container. Fermentation takes about seven days.

Pickled beetroots

In Russia beetroot is pickled by cleaning, slicing and placing in a container with salt. Due to the high sucrose level, dextrans are produced giving the product a slimy texture.

Lamoun Makbous (pickled lemons)

Pickled lemons are popular in Asia. In west Asia and north Africa they are known as *lamoun makbous* and *msir*. Lemons are washed in clean water, sliced and covered in salt. After at least 24 hours, they are drained and mixed with oil and spices.

Brined Fruit and Vegetable Pickles

For brine pickling any variety of common salt is suitable as long as it is pure. Impurities or additives can cause problems:

— Chemicals to reduce caking should not be used as they make the brine cloudy.

— Lime impurities can reduce the acidity of the final product and reduce the shelf life of the product.

— Iron impurities can result in the blackening of the vegetables.

— Magnesium impurities impart a bitter taste.

— Carbonates can result in pickles with a soft texture.

Green Mango Pickle

Location of production

Mango pickle is a very popular pickle in many Asian, African and Latin American countries. It is a major product of India, Pakistan and Bangladesh and it is estimated that the annual production of mango pickle in South Africa is over 10,000 tons.

Product description

Green mango pickle is a hot, spicy pickle with a sour taste. It is eaten as a condiment. Preservation is caused by a combination of salt, increased acidity and to a small extent the spices. It is known as *burong mangga* and *dalok* in the Philippines.

Preparation of the raw material

The fresh, fully mature, firm but unripe mangoes must be carefully selected to ensure a good quality product. The best pickles are obtained from fruit at early maturity when the fruit has reached almost maximum size. Riper fruit results in pickles with a fruity odour and lacking the characteristic and predominant green mango flavour.

The green mangoes need to be inspected and any damaged fruit rejected. The fruit is washed in clean water and drained. After draining, the fruit is cut. Sharp knives with preferably stainless blades should be used. Iron or copper equipment should be avoided. A single stroke should be used during the cutting process to ensure minimum damage and avoiding mushiness in the final product.

Processing

The sliced mangoes are soaked in brine solution. Sodium metabisulphite (1000 ppm) and 1% calcium chloride are added. The containers are stored until the mangoes are pickled. The brine is then drained off and spices are mixed with the mango slices.

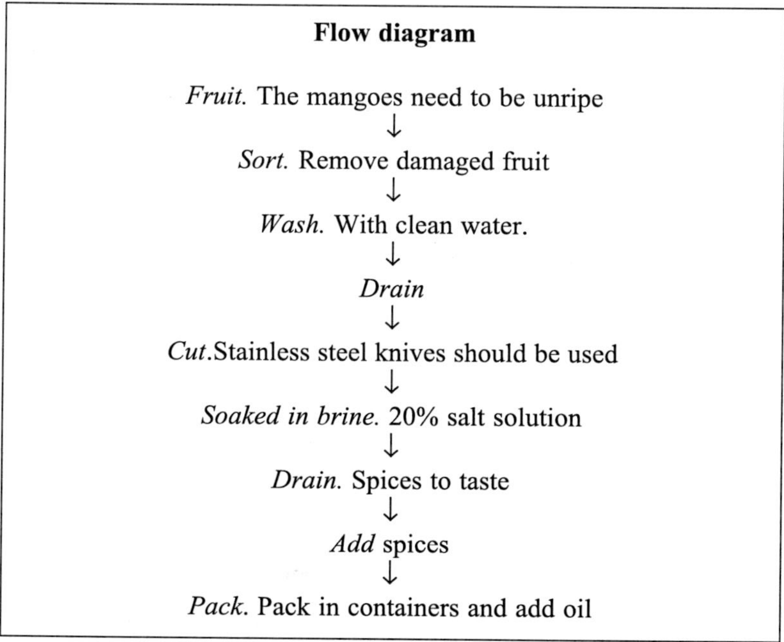

Packaging and storage

The mixture is then packed and oil added onto the surface of the mixture. The mangoes should be firmly pressed down in the container. Good quality vegetable oil such as sunflower oil should be used and finely ground chilli powder can be added to the oil for flavour and colour. Mango pickle can be packed in small polythene bags and sealed or in clean jars and capped. Mango pickle keeps well if stored in a cool place. If it is processed well, it can be kept for several months. Due to the high acid level of the final product, the risk of food poisoning is low.

Lime Pickle

Location of production

Lime pickles are produced in Asia, Latin America and Africa. They are particularly popular in India, Pakistan and North Africa.

Product description

Lime pickle is made from salted pieces of lime packed in a salty, spicy liquor, like a semi-solid gravy. It is brownish red and the lime peels are yellow or pale green with a sour and salty taste. It is eaten as a condiment with curries or other main meals. If processed well, the product can be kept for several months.

Preparation of raw materials

The limes need to be selected and prepared. Only fully ripe limes without bruising or damage should be used. All the limes need to be washed in potable cold water and drained. The limes are dipped in hot water (60-65^{0}C) for about five minutes. They are then cut into pieces in order to expose the interior and allow salt to be absorbed more quickly. All spices should be of good quality and free of mould.

Processing

The prepared limes are covered with a brine solution. This causes water to be drawn out of the pieces by osmosis. It is important to ensure that the surface is covered with juice, and leave for 24 hours. If necessary, the fruits should be pressed down to hold them below the liquid. Once the limes have been placed in the brine, there is a rapid development of micro-organisms and fermentation begins.

After fermentation the limes are dried in the sun until the skin becomes brown.

Flow diagram

Sort. Select ripe (but not overripe) healthy lime fruits

↓

Wash

↓

Heat. Dip in hot water (60-65^{0}C) for about five minutes

↓

Cut. Cut into four pieces or alternatively cut into smaller, uniform-sized pieces

↓

Brine. Ensure that the surface is covered with juice

↓

Dry. Dry in the sun for 2-3 days

↓

Mix spices. To local preference

↓

Pack

↓

Store. In a cool place, away from sunlight

Packaging and storage

The limes are mixed with spices and oils according to local taste and tradition. Lime pickle can be packed in small polythene bags and sealed or in clean jars and capped. Lime pickle keeps well if stored in a cool place. Due to the high acid level of the final product, the risk of food poisoning is low.

Kimchi (pickled cabbage)

Location of production

Kimchi is probably the most important processed food product in Korea. It is an essential dish, eaten at most mealtimes. Production is estimated at over one million tons, mainly at household level. Daily consumption is estimated at 150 to 250 grams per person.

Product description

Kimchi is a general name for a range of closely related fermented products. It is similar to Sauerkraut in Europe and the United States. There are numerous variations of *kimchi* depending on the production technique. The main pickled cabbage *kimchis* are *tongbaechu-kimchi tongkimchi* and *bossam-kimchi.*

Preparation of raw materials

Appropriate cultivars of Chinese cabbage, with light-green coloured soft leaves and compact structures with no defects, are required for production of *kimchi.* After removing outer leaves and roots from the cabbage, it is cut into small pieces.

Processing

The prepared cabbage is placed in a salt solution (8-15%) for two to seven hours in order increase the salt content of the cabbage to between 2.0-4.0% (w/w). It is then rinsed several times with fresh water and drained to remove extra water by centrifugation or by allowing to stand.

Kimchi fermentation is carried out by various micro-organisms present in the raw materials and ingredients used in the preparation of*kimchi*. Among the two hundred bacteria isolated form kimchi, the important micro-organisms in *kimch*i fermentation are known to be*Lactobacillus plantarum, L. Brevis, Streptococcus faecalis, Leuconostoc mesenteroides* and *Pediococcus pentosaceus.*

After fermentation, the product can be left to mature for several weeks if refrigeration is available. If stored under warm conditions, the*kimchi* deteriorates rapidly.

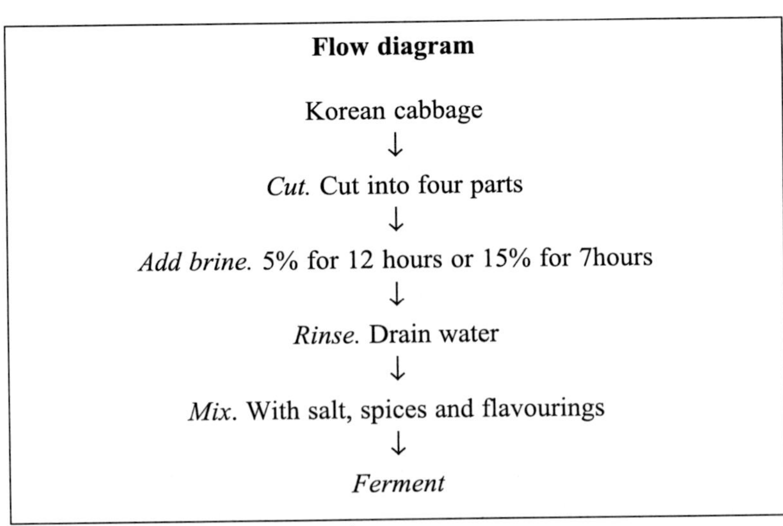

Green olives

Olives are a brine fermented product which undergo an essential pre-treatment with lye to remove substances which are toxic to bacteria and would, if left in the olives, prevent fermentation.

Green olives are placed in a 2% sodium hydroxide (lye) solution at 21^0 to 24^0C until the lye penetrates the flesh. Cold water is added to the solution, which, dilutes the mixture until the lye is completely removed. The lye treatment is necessary to remove a bitter glucoside compound (oleuropein) from the outer tissues of the olive. Oleuropein is highly toxic to bacteria and therefore needs to be removed in order for a fermentation to take place. After the glucoside has been removed, the olives are placed in barrels with a 1 to 10% brine solution and allowed to undergo a spontaneous fermentation. The optimum fermentation temperature is 24^0C. The fermentation period usually takes between two and three months. Once fermentation is complete, the olives are packed in airtight jars and sterilised which produces a good quality product with a long storage life.

Black olives

When ripe olives are used, they are placed in a 5 to 7% brine solution to soften the outer tissue before lye is added. This pre-treatment allows the lye to penetrate the fruit more easily. In ripe fruit, the glucoside is situated more deeply within the tissue. After soaking in brine, the lye solution (0.7 to 2.0%) is added and, after soaking, the olives are washed clean. They are then packed in barrels with a 2 to 5% brine solution and allowed to undergo fermentation for two to six weeks, depending on the external temperature. Unlike immature olives, the ripe one are exposed to air during fermentation. This is to oxidise the polyphenols in the tissue to a black colour, which is dependent upon oxygen and the small amounts of sodium hydroxide which are left on the olives. The finished product is packed in a 3 to 5% brine solution and then sterilised. *Bacillus subtilis* may produce pectolytic enzymes which result in a soft product. If the wash water used for removing the lye is below 60^0C, many species of undesirable bacteria can survive which result in a low quality product.

Jack-fruit pickle

Young green jack fruit is pickled in India and Sri Lanka. Young green jack fruit are peeled and cut into 1.2 to 1.8 cm thick slices. The slices are placed in a container and covered in an 8% common salt solution. They are weighed down to keep them submerged in the brine. The brine solution is increased by 2% each day until it reaches 15%. The slices are then left for 8-10 days in the brine. Vinegar and spices are added prior to packaging.

Pickled Radish

A number of pickled radish products are produced in Korea. These include: *kaktugi, tongchimi,chonggak-kimchi, seokbakji, yolmu-kimchi dan moogi kach doo ki gactuki and mootsanj*i.

Pickled cucumber

A variety of brine pickled cucumber products are made around the world. *Oi sobagi* and *oiji* are made in Korea. In Egypt cucumbers are pickled by soaking in brine to produce *torshi khiar*.

Pickled leafy vegetables

There are many other brine pickled leafy vegetables around the world. For instance:

— *Pak-sian-dong*: This is a popular pickled leafy vegetable *Pak Sian* (*Gynadropsis pentaphylla*) in Thailand. The fresh vegetable is cleaned and wilted in the sun for one to two hours. It is then placed in brine and fermented for two to three days at room temperature

— *Sayur asin* fermented wilted mustard cabbage (*Brassica juncea*) from Indonesia. It is also known as *kiam chai* in Thailand and*kiam chaye* in Malaysia.

Other pickled vegetables and fruits

— *Naw-mai-dong*: pickled bamboo shoots (*Bambusa glaucescens*) from Thailand

— *Hom-dong*: pickled red onions from Thailand

— Jeruk: pickled vegetables including ginger and papaya from Malaysia

— Pickled carrots and turnips are produced in Asia and Africa. They are known as *hua-chai po* in Thailand and *tai tan tsoi* in China.

— *Nukamiso-zuke*: vegetables fermented in rice bran, salt and water in Japan.

— Bananas are pickled in the West Indies

— Fermented sweet peppers (*torshi felfel*) are produced in west Asia and Africa.

— Cauliflower stalks are fermented to produce *achar tandal* in India.

— Aubergines (*torshi betingen*) are pickled in west Asia.

Variety of Pickles around the World

South Asia

India has a large variety of pickles (known as Achar in Punjabi and Hindi, Uppinakaayi in Kannada, Lonacha in Marathi, Oorukai in Tamil, which are mainly made from mango, lime, Indian gooseberry (amla), chilli, vegetables such as egg plants, carrots, cauliflower,tomato, bitter gourd & green tamarind etc., ginger, garlic, onion and citron. These fruits/vegetables are generally mixed with some other ingredients i.e. salt, spices, vegetable oils and is set to mature. A special variety of mango pickle prepared in India's Andhra Pradesh State called Avakkaya is very popular. This pickle, often described as king of all pickles, is a fine blend of cut mango pieces, red chilli powder, mustard seed powder, salt and sesame oil each ingredient used in specified proportion all raw and allowed to mature for few weeks in porcelain containers in hot summer.

In Iran, pickles (called torshi in Persian) are commonly made from turnips, peppers, carrots, green olives, cucumbers, cabbage, lemons, and cauliflower.

In Pakistan, pickles are known locally as Achaar (in Urdu) and come in a variety of flavours. Amongst some of the most popular is the traditional mixed Hyderabadi pickle, a common delicacy and staple prepared from an assortment of fruits (most notably mangos) and vegetables blended with selected spices.

East Asia

Indonesian and Malaysian pickles, acar are typically made out of cucumber, carrot, bird's eye chilies, and shallots, these items being seasoned with vinegar, sugar and salt. Fruits, such as papaya and pineapple are also sometimes pickled. In the Philippines, achara is primarily made out of green papaya, carrots, and shallots, with cloves of garlic and vinegar. In Vietnam, vegetable pickles are called dua mui (salted vegetables) or dua chua ("sour vegetables"). In Sri Lanka, achcharu is traditionally prepared out of carrots, onions, and ground dates. Mixed with mustard powder, ground pepper, crushed ginger, garlic and vinegar, these items are seasoned in a clay pot.

China is home to a huge variety of pickled vegetables, including radish, baicai (Chinese cabbage, notably suan cai, la bai cai, pao cai, and Tianjin preserved vegetable), zha cai, chili pepper and cucumber, among many others.

Japanese tsukemono (pickled foods) include takuan (daikon), umeboshi (ume plum), gari & beni shoga (ginger), turnip, cucumber, and Chinese cabbage.

The Korean staple kimchi is usually made from pickled cabbage and radish, but is also made from green onions, garlic stems, chives and a host of other vegetables. Kimchi is popular throughout East Asia. Jangajji is another example of pickled vegetables.

Middle East

In Arab countries, pickles (called mekhallel in Arabic) are commonly made from turnips, peppers, carrots, green olives, cucumbers, beetroot, cabbage, lemons, and cauliflower.

Asia Minor

Turkish pickles, called tursu, are made out of vegetables, roots, and fruits such as peppers, cucumber, Armenian cucumber, cabbage, tomato, eggplant (aubergine), carrot, turnip, beetroot, green almond, baby watermelon, baby cantaloupe, garlic, cauliflower, bean and green plum. A mixture of spices flavor the pickles.

Eastern Europe

Coriander seeds are one of the spices popularly added to pickled vegetables in Europe.

Romanian pickles are made out of beetroot, cucumbers, green tomatoes, carrots, cabbage, bell peppers, melons, mushrooms, turnips, celery and cauliflower. Meat, like pork, can also be preserved in salt and lard.

In Albania, Bulgaria, Serbia, Macedonia and Turkey, mixed pickles, known as turshi or turshu form popular appetizers, which are typically eaten with rakia. Pickled green tomatoes, cucumbers, carrots, bell peppers, peppers, eggplants, and sauerkraut are also popular.

Polish traditional pickles are cucumbers and cabbage, but other pickled fruits and vegetables, including plums, pumpkins and mushrooms are also common.

Russian pickled items include beets, mushrooms, tomatoes, cabbage, cucumbers, ramsons, garlic, eggplant (which is typically stuffed with julienned carrots), custard squash, and watermelon.

Western Europe

In Britain, pickled onions and pickled eggs are often sold in pubs and fish and chip shops. Pickled beetroot, walnuts, and gherkins, and condiments such as Pickle and piccalilli are typically eaten as an accompaniment to pork pies and cold meats, sandwiches or a ploughman's lunch. Other popular pickles in the UK are pickled mussels, cockles, red cabbage, mango chutney, sauerkraut, and olives.

Southern Europe

An Italian pickled vegetable dish is giardiniera, which includes onions, carrots, celery and cauliflower. Many places in southern Italy, particularly in Sicily, pickle eggplants and hot peppers.

Northern Europe

Pickled herring, rollmops, and salmon are popular in Scandinavia. Pickled cucumbers and red garden beets are important as condiments for several traditional dishes. Pickled capers are also common in Scandinavian cuisine.

North America

In the United States and Canada, pickled cucumbers (most often referred to simply as "pickles" in Canada and the United States), olives, and sauerkraut are most popular, although pickles popular in other nations (such as the pickled tomatoes commonly offered in New York City delicatessens) are also available. Giardiniera, a mixture of pickled peppers, celery and olives, is a popular condiment in Chicago and other cities with large Italian-American populations, and is often consumed with Italian beef sandwiches. Pickled eggs are common in the Upper Peninsula of Michigan. Pickled herring is available in the Upper Midwest. Pennsylvania Dutch Country has a strong tradition of pickled foods, including chow-chow and red beet eggs. In the Southern United States, pickled okra and watermelon rind are popular, as are deep-fried pickles and pickled pig's feet, chicken eggs, quail eggs and pickled sausage. In Mexico, chili peppers, particularly of the Jalapeño and serrano varieties, pickled with onions, carrots and herbs form common condiments.

References

Chou, Lillian. 2003. "Chinese and Other Asian Pickles". *Flavor and Fortune.* Institute for the Advancement of the Science And Art Of Chinese Cuisine. Retrieved 6 December 2012.

McGee, Harold. 2004. *On Food and Cooking: The Science and Lore of the Kitchen.* New York: Scribner, pp. 291–296.

Zeldes, Leah A. 2009. "Eat this! Southern-fried dill pickles, a rising trend". *Dining Chicago*. Chicago's Restaurant & Entertainment Guide, Inc.. Retrieved 2010-08-02.

Islami, F. 2011. "Pickled vegetables and the risk of oesophageal cancer: a meta-analysis". *British Journal of Cancer*. Retrieved. 2011-08-16.

7

Food Irradiation

Food irradiation is the treatment of food by a certain type of energy. The process involves exposing the food, either packaged or in bulk, to carefully controlled amounts of ionizing radiation for a specific time to achieve certain desirable objectives as will be detailed later in the text. The process cannot increase the normal radioactivity level of the food, regardless of how long the food is exposed to the radiation, or how much of an energy "dose" is absorbed. It can prevent the division of microorganisms which cause food spoilage, such as bacteria and moulds, by changing their molecular structure. It can also slow down ripening or maturation of certain fruits and vegetables by modifying/altering the physiological processes of the plant tissues.

Trends in Food Irradiation

Alongside traditional methods of processing and preserving food, the technology of food irradiation is gaining more and more attention around the world. Although regarded as a new technology by some individuals, research on food irradiation dates back to the turn of the century with the first USA and British patents being issued in 1905 for the use of ionizing radiation to kill bacteria in food. Today, health and safety authorities in over 40 countries have approved irradiation of over 60 different foods, ranging from spices to grains to deboned chicken meat, to beef, to fruits and vegetables. As of August 1999, over 30 countries are irradiating food for commercial purposes. There are approximately 60 irradiation facilities being used for this purpose with more under construction or at the planning stage.

Decisions in these and other countries to irradiate food have been influenced by the adoption, in 1983, of a worldwide standard covering irradiated foods. The standard was adopted by the Codex Alimentarius Commission, a joint body of the Food and Agriculture Organization of the United Nations (FAO) and the World Health Organization (WHO), responsible for issuing food standards to protect consumer health and facilitate fair practice in food trade, representing more than 150 governments. The Codex General Standard for food irradiation was based on the findings of a Joint Expert Committee on Food Irradiation (JECFI) convened by the FAO, WHO, and the International Atomic Energy Agency (IAEA). The JECFI has evaluated available data in 1964, 1969, 1976, and 1980. In 1980, it concluded that "the irradiation of any food commodity" up to an overall average dose of 10 kGy "presents no toxicological hazard" and requires no further testing. It stated that irradiation up to 10 kGy "introduces no special nutritional or microbiological problems" in foods. In September 1997 a Study Group was jointly convened by the WHO, FAO and IAEA to evaluate the wholesomeness of food irradiated with doses above 10 kGy. This Study Group concluded that there is no scientific basis for limiting absorbed doses to the upper level of 10 kGy as currently recommended by the Codex Alimentarius Commission. Food irradiation technology is safe to such a degree that as long as the sensory qualities of food are retained and harmful microorganisms are destroyed, the actual amount of ionizing radiation applied is of secondary consideration.

Interest in the irradiation process is increasing because of persistently high food losses from infestation, contamination, and spoilage; mounting concerns over food-borne diseases; and growing international trade in food products that must meet strict import standards of quality and quarantine, all areas in which food irradiation has demonstrated practical benefits when integrated within an established system for the safe handling and distribution of food. In addition, with increasingly restricted regulations or complete prohibition on the use of a number of chemical fumigants for insect and microbial control in food, irradiation is an effective alternative to protect food against insect damage and as a quarantine treatment of fresh produce.

The FAO has estimated that worldwide about 25% of all food production is lost to insects, bacteria and rodents after harvesting. The use of irradiation alone as a preservation technique will not solve all the problems of post-harvest food losses, but it can play an important role in cutting losses and reducing the dependence on chemical pesticides. Many countries lose vast amounts of grain because of insect infestation and moulds. For roots and tubers, sprouting is the major cause of losses. Several countries, including Bangladesh, Chile, China, Hungary, Japan, Republic of Korea and Thailand are irradiating one or more food products (grains, potatoes, spices, dried fish, onions, garlic, etc.) to control food losses on a commercial basis.

Foodborne diseases pose a widespread threat to human health and they are an important cause of reduced economic productivity even in advanced countries which have modern food processing and distribution systems. Although the amount of foodborne disease caused by pathogenic bacteria in the United States is not known with accuracy, it was estimated in 1994 by a task force of the Council for Agricultural Science and Technology (CAST) that the number of cases likely range from 6.5 million to 33 million annually and that deaths may be as high as 9,000 annually. The United States Department of Agriculture's (USDA) Economic Research Service estimates that diseases caused by E. coli O157:H7 due to consumption of insufficiently cooked ground beef result in US $200 million to $440 million in annual medical costs and productivity losses. In developing countries, diseases caused by parasites such as Taenia solium and Trichinella spiralis constitute a major problem, and together with bacterial foodborne diseases, account for hundreds of millions of cases per year.

The relatively low doses of radiation needed to destroy certain bacteria in food can be useful in controlling foodborne disease. Considerable amounts of frozen seafoods and frog legs, as well as dry food ingredients, are irradiated for this purpose in Belgium, France and the Netherlands. Electron beam irradiation of blocks of frozen mechanically deboned, poultry meat is carried out industrially in France. Spices are being irradiated (instead of being fumigated) in many countries including Argentina, Belgium, Brazil, Canada, China, Denmark, Finland, France, Hungary, Indonesia, Israel, Mexico, the Netherlands, Norway, Republic of Korea, South Africa, the United Kingdom and the USA. The volume of irradiated spices and dried vegetable seasonings globally has increased significantly in recent years to over 60,000 tonnes in 1997.

Trade in food products is a major factor in regional and international commerce, and markets are growing. The inability of countries to satisfy each other's quarantine and public health regulations is a major barrier to trade. For example, not all countries allow importation of chemically-treated fruit. Moreover, major importing countries, including the USA and Japan, have banned the use of and the import of produce treated with certain fumigants identified as health hazards. During 1996, the United States Department of Agriculture (USDA) issued a new policy to allow importation of fresh fruits and vegetables treated by radiation against fruit flies. The problem is most acute for developing countries whose economies are still largely based on food and agricultural production and the revenues from export. Radiation processing offers these countries an alternative to fumigation and some other treatments.

Each year a few hundred thousand tonnes of food products and ingredients are irradiated worldwide. This amount is small in comparison to the total volumes of processed foods and not many of these irradiated food products enter international commerce.

One factor influencing the speed with which food irradiation is being adopted is public understanding and acceptance of the process. Contrary to earlier estimates it has been demonstrated that when irradiated foods are available, consumers have purchased them because of their satisfaction with product quality and safety. It is normal to seek reassurance as to the safety and effectiveness of any new process or technology. Therefore, it is hoped that this revised publication will help address concerns and correct myths about food irradiation.

Scientific and Technical Terms

The type of radiation used in processing materials is limited to radiations from high energy gamma rays, X-rays and accelerated electrons. These radiations are also referred to as ionizing radiations because their energy is high enough to dislodge electrons from atoms and molecules and to convert them to electrically-charged particles called ions.

Gamma rays and X-rays, like radiowaves, microwaves, ultraviolet and visible light rays, form part of the electromagnetic spectrum and occur in the short-wavelength, high-energy region of the spectrum and have the greatest penetrating power. They have the same properties and effects on materials, their origin being the main difference between them. X-rays with varying energies are generated by machines. Gamma rays with specific energies come from the spontaneous disintegration of radionuclides.

Naturally occurring and man-made radionuclides, also called radioactive isotopes or radioisotopes, emit radiation as they spontaneously revert to a stable state. The time taken by a radionuclide to decay to half the level of radioactivity originally present is known as its half-life, and is specific for each radionuclide of a particular element. The becquerel (Bq) is the unit of radioactivity and equals one disintegration per second.

Only certain radiation sources can be used in food irradiation. These are the radionuclides cobalt-60 or cesium-137; X-ray machines having a maximum energy of five million electron volts (MeV) (an electron volt is the amount of energy gained by an electron when it is accelerated by a potential of one volt in a vacuum); or electron accelerators having a maximum energy of 10 MeV. Energies from these radiation sources are too low to induce radioactivity in any material, including food.

The radionuclide used almost exclusively for the irradiation of food by gamma rays is cobalt-60. It is produced by neutron bombardment in a nuclear reactor of the metal cobalt-59, then doubly encapsulated in stainless steel "pencils" to prevent any leakage during its use in an irradiator. Cobalt-60 has a half-life of 5.3 years, the gamma rays produced are highly penetrating and can be used to treat full boxes of fresh or frozen food.

Cesium-137 is the only other gamma-emitting radionuclide suitable for industrial processing of materials. It can be obtained by reprocessing spent, or used, nuclear fuel elements and has a half-life of 30 years. However, there is no supply of commercial quantities of cesium-137. Cobalt-60 has therefore become the choice for gamma radiation source; over 80% of the cobalt-60 available in the world market is produced in Canada. Other producers are the Russian Federation, the People's Republic of China, India and South Africa.

High energy electron beams can be produced from machines capable of accelerating electrons to near the speed of light by means of a linear accelerator. Since electrons cannot penetrate very far into food, compared with gamma radiation or X-rays, they can be used only for treatment of thin packages of food and free flowing or falling grains. X-rays of various energies are produced when a beam of accelerated electrons bombards a metallic target. Although X-rays have good penetrability into food, the efficiency of conversion from electrons to X-rays is generally less than 10%, and this has hindered the use of this type of radiation source so far.

Radiation dose is the quantity of radiation energy absorbed by the food as it passes through the radiation field during processing. It is measured using a unit called the Gray (Gy). In early work the unit was the rad (1 Gy = 100 rads; 1 kGy = 1000 Gy). International health and safety authorities have endorsed the safety of irradiation for all foods up to a dose level of 10,000 Gy (10 kGy). Recent evaluation of an international expert study group appointed by FAO, IAEA and WHO showed that food treated according to good manufacturing practices (GMPs) at any dose above 10 kGy is also safe for consumption, making irradiation parallel to heat treatment of food. In terms of energy relationships, one gray equals one joule of energy absorbed per kilogram of food being irradiated.

The maximum dose of 10 kGy recommended by the Codex General Standard for Irradiated Foods is equivalent to the heat energy required to increase the temperature of water by 2.4°C. Irradiation is often referred to as a "cold pasteurization" process as it can accomplish the same objective as thermal pasteurization of liquid foods, for example milk, without any substantial increase in product temperature.

Table 1. Units of Radiation Dose and Radioactivity

	Absorbed dose	*Radioactivity*
Unit	gray (Gy)	becquerel (Bq)
Definition	1 Gy = 1 J/kg	1 Bq = 1 disintegration/sec
Former unit	rad	curie (Ci)
Conversion	1 rad = 0.01 Gy	1 Ci = 3.7 X 10^{10}Bq = 37 GBq
	1 krad = 10 Gy	1 kCi = 37 TBq
	1 Mrad = 10 kGy	1 mCi = 37 PBq

Benefits of Food Irradiation

Finding ways to prevent the deterioration of food and control infection by microorganisms has been a major preoccupation of man over the centuries. Controls such as refrigeration or pasteurization are now commonplace, and it is expected that one day the technique of food irradiation will also be widely used. Food irradiation can offer a wide range of benefits to food industry and the consumer. From a practical point of view, there are three general application and dose categories that are referred to when foods are treated with ionizing radiation:

— *Low-dose irradiation* – up to 1 kGy (sprout inhibition; delay of ripening; insect disinfestation; parasite inactivation).

— *Medium-dose irradiation* – 1 to 10 kGy (reduction in numbers of spoilage microorganisms; reduction in numbers or elimination of non-spore-forming pathogens, i.e. disease causing microorganisms).

— *High-dose irradiation* – above 10 kGy (reduction in numbers of microorganisms to the point of sterility).

Perhaps the most important application of this method of food preservation is to ensure the hygienic quality of solid or semi-solid foods, especially those of animal origin, through inactivation of foodborne pathogens. This application is comparable to thermal pasteurization of liquid foods, for example milk, which is effective and widely applied but is unsuitable for foods such as poultry, meat and seafood that are to be marketed in raw form.

Reduction of Pathogenic Microorganisms

The incidence of foodborne disease arising from the consumption of food contaminated with pathogenic microorganisms is increasing, and there is a heightened public awareness of the health threat posed by pathogens in or on food. Among these, Escherichia coli O157:H7, Salmonella, Campylobacter jejuni, Listeria monocytogenes, and Vibrio are of primary concern from a public health standpoint because of the severity of the illnesses and/or because of the higher number of outbreaks and individual cases of foodborne illness associated with these pathogens. Of these food poisoning bacteria, Salmonella and C. jejuni are usually associated with poultry. E. coli O157:H7 has also been linked to major foodborne disease outbreaks through many sources including meat and dairy products in the United Kingdom, hamburger meat, apple juice and water in the USA, and vegetables in Japan. Listeria monocytogenes has been associated with dairy products, processed meats and other foods having a relatively long shelf-life under refrigeration. Vibriospp. in turn have been the causitive agents in world cholera pandemics and of many outbreaks of disease caused by consumption of raw mollusks.

Adherence to good manufacturing practice (GMP) is obviously essential but this alone may not be sufficient to reduce the number of food poisoning outbreaks. Pathogens such as those mentioned previously are sensitive to low levels of ionising radiation. As the irradiation dose increases more microorganisms are affected but a higher dose, although not creating any harmful products, may simultaneously introduce changes in sensory qualities and therefore a balance must be attained between the optimum dose required to achieve a desired objective and that which will minimise any sensory changes. This is also the case for thermal-pasteurization– following his research on spoilage in wine, Louis Pasteur recommended a level of heat treatment that was sufficient to inactivate spoilage microorganisms but not high enough to destroy the quality or character of the products processed. He therefore determined the minimum processing required to attain the desired objective without impairing the product's overall acceptability.

With fresh poultry carcasses, irradiation up to a dose of 2.5 kGy will virtually eliminate Salmonella and Campylobacter under proper production conditions. The same dose of irradiation destroys E. coli O157:H7, a highly virulent bacteria which can lead to illness and death, and which is estimated to cause 20,000 infections and 250 deaths in the USA annually. Irradiation is currently the only known method to inactivate these pathogens in raw and frozen food.

Frog legs can be heavily contaminated by Salmonella and other pathogens, and irradiation provides an effective means of decontamination. French importers have routinely irradiated this product for a number of years, and irradiated frozen frog legs can be purchased in most French food markets. Eggs and egg products are often contaminated with Salmonella and have been the subject of many food irradiation studies. Early work in the United Kingdom showed that frozen egg and dried egg could be irradiated at doses of up to 5 kGy without quality loss and that this dose provided sufficient hygienic protection. More recent work suggests 2 kGy as the most suitable dose for inactivation of Salmonella in egg powder; at the same time preserving the sensory and technological properties.

Seafood, especially shellfish, is often contaminated with pathogenic organisms such as Salmonella, Vibrio parahaemo-lyticus, and Shigella. Consumption of raw and inadequately cooked shellfish is considered to present unacceptable risk factors. Nevertheless, many people do eat raw shellfish such as oysters and clams. In frozen shrimp, reduction of pathogens to a safe level requires a dose of about 3 kGy for inactivating Vibrio spp., Salmonella spp. and Aeromonas hydrophila.

It is well known in some countries that fresh pork meat must be cooked thoroughly because it may contain Trichinella spiralis, a parasite which may cause illness and death. The larvae of this parasite can be rendered non-infective by irradiation with a minimum dose of 0.3 kGy. Pork treated in this manner is known as "trichina-safe". Another pork parasite Toxoplasma gondii can also be inactivated with a minimum dose of 0.5 kGy.

Decontamination

Spices, herbs and vegetable seasonings are valued for their distinctive flavours, colours and aromas. However, they are often heavily contaminated with microorganisms because of the environmental and processing conditions under which they are produced. Therefore, before they can be safely incorporated into other food products, the microbial load should be reduced. Because heat treatment can cause significant loss of flavour and aroma, a "cold process", such as irradiation, is ideal. Until recently, most spices and herbs were fumigated, usually with sterilizing gases such as ethylene oxide to destroy contaminating microorganisms. However, the use of ethylene oxide was prohibited by an European Union (EU) directive in 1991 and has been banned in a number of other countries because it is a carcinogen. Irradiation has since emerged as a viable alternative and its use results in cleaner, better quality herbs and spices compared to those fumigated with ethylene oxide. Irradiation of spices on a commercial scale is practised in over 20 countries and global production has increased significantly from about 5,000 tonnes in 1990 to over 60,000 tonnes in 1997. In the USA alone over 30,000 tonnes of spices, herbs and dry ingredients were irradiated in 1997 as compared to 4,500 tonnes in 1993.

Extension of Shelf-life

The shelf-life of many fruits and vegetables, meat, poultry, fish and seafood can be considerably prolonged by treatment with combinations of low-dose irradiation and refrigeration that do not alter flavour or texture. Many spoilage microorganisms, such as Pseudomonas spp., are relatively sensitive to irradiation. For example, a dose of 2.5 kGy applied to fresh poultry carcasses processed according to good manufacturing practices (GMPs) will be enough to eliminate Salmonella, and will also kill many, but not all, spoilage bacteria. This will double meat shelf-life, provided it is kept below 5°C.

Extension of the very short shelf-life of many commercially important plant commodities is highly desirable, and in some cases, critical. Exposure to a low dose of radiation can slow down the ripening of some fruits, control fungal rot in some others and maturation in certain vegetables, thereby extending their shelf-life. For example, ripening in bananas, mangoes,and papayas can be delayed by irradiation at 0.25 to 1 kGy. Strawberries are frequently spoiled by Botrytis mould. Treatment with a dose of 2 to 3 kGy followed by storage at 10°C can result in a shelf-life of up to 14 days, but the extension obtained depends on the initial quality of the fresh food, which should be as good as possible. Irradiation of mushrooms at 2 to 3 kGy inhibits cap opening and stem elongation. Shelf-life extension can be increased at least two-fold by irradiation and subsequent storage at 10°C, and even longer when stored at a lower temperature compared with non-irradiated mushrooms.

Not all fruits and vegetables are suitable for irradiation because undesirable changes in colour or texture, or both, limit their acceptability. Also, different varieties of the same fruit or vegetable may respond differently to irradiation. The time of harvest and the physiological state also affects the response of fruits and vegetables to irradiation. For example, if strawberries are irradiated before they are ripe, the red colour does not develop satisfactorily. For delaying ripening in fruits it is important to irradiate them before ripening starts.

At high doses of irradiation (>25 kGy), foods which are preheated to inactivate enzymes can be commercially sterilized such as occurs in canning. The sterilized products can be stored at room temperature almost indefinitely. Radiation-sterilized foods are given to hospital patients who have immune system deficiencies and must therefore have a sterile diet. Irradiation sterilized products are also eaten by astronauts in the NASA space shuttle programme because of their superior quality, safety and variety, in preference to foods treated by other preservation techniques. Limited commercial-scale sterilization of various ready-to-eat foods by high dose irradiation has been carried out in South Africa during the past 10 years to serve military personnel and outdoor enthusiasts such as campers, yachters and hikers. In total, more than two million light weight food packs (weighing 150 g each) have been produced during this period.

Disinfestation

The chief problem encountered in preservation of grains and grain products is insect infestation. Most of the pests of concern, e.g. beetles, moths, weevils and others, are not quarantine insects, but they cause extensive damage to stored products. Irradiation has been shown to be an effective pest control method for these commodities and a good alternative to methyl bromide, the most widely used fumigant for insect control, which is being phased out globally because of its ozone depleting properties. Unlike methyl bromide, irradiation is not an ozone depleting substance and unlike phosphine, the other major fumigant used to control grain pests, irradiation is a fast treatment and its efficacy is not temperature dependent. Irradiation can kill or control phos-phine-resistant pests. The dosage required for insect control is reasonably low, in the order of 1 kGy or less. Disinfestation is aimed at preventing losses caused by insects in stored grains, pulses, flour, cereals, coffee beans, dried fruits, dried nuts, and other dried food products including dried fish. Proper packaging is required, however, for irradiated products to prevent insect reinfestation.

Radiation disinfestation can facilitate trade in fresh fruits, such as citrus, mangoes, and papayas which often harbour insect pests of quarantine importance. Insects are easily distributed by international tradc in such fruits and also by tourism. To prevent or minimize this risk, many countries prohibit importation of such fruits or require quarantine treatment of imported fruits. These measures can create significant barriers to international trade and the free flow of plants and plant products, but they are fully justified from the receiving country's point of view. The occurrence of fruit flies, such as the Mediterranean, Oriental, Mexican or Caribbean fruit flies, has repeatedly disrupted trade among countries and between states within large countries, for example Australia and the USA. A number of quarantine treatments permitted in the past have recently been banned, fumigation with ethylene dibro-mide being the most prominent example. It has been demonstrated that low dosages of ionizing radiation, between 0.15 and 0.3 kGy, will very effectively control fruit fly and other insect problems. This makes the use of irradiation for quarantine treatment a very practical possibility. In 1996, the United States Department of Agriculture (USDA)/ Animal and Plant Health Inspection Service (APHIS) issued a Notice of Policy accepting irradiation as a quarantine treatment against major species of fruit fly regardless of commodities. Subsequently, in 1997 a final rule was issued by the USDA/APHIS for the irradiation of papayas, carambola, and litchi as a phyto-sanitary treatment. The rule allows interstate movement of these commodities from Hawaii to the USA mainland and permits treatment either in Hawaii or in non-fruit fly supporting areas of the mainland USA. Small commercial scale irradiation of fruits from Hawaii has been carried out under special permission of the USDA/APHIS since 1995. Such irradiated fruits have been marketed successfully in the USA.

Inhibition of Sprouting

In order to provide consumers with a year-round supply of potato tubers, onion bulbs, yams and other sprouting plant foods, storage over many months is necessary unless shipments from other climatic zones, usually at a much higher price, can replace local production during off-season. Such long-term storage is possible with the aid of refrigeration, which is costly, particularly in subtropical and tropical regions. For many of these crops, the desired inhibitory effects can also be obtained using chemical sprout inhibitors such as maleic hydrazide, propham, or chloropropham. These chemicals, however, are either not effective under tropical conditions or leave residues in the produce, and for health reasons they are considered by some to be harmful. Thus many countries have prohibited their use.

A very low radiation dose of 0.15 kGy or less, inhibits sprouting of products such as potatoes, yams, onions, garlic, ginger, and chestnuts. It leaves no residues and allows storage at higher temperatures. Irradiation of potatoes, stored at higher temperatures (10°-15°C), have better processing quality. Commercial processing of irradiated potatoes has been carried out in Japan since 1973.

Neither irradiation nor any other food treatment can reverse the spoilage process and make bad food good. If food already looks, tastes or smells bad – signs of spoilage – before irradiation, it cannot be "saved" by any treatment including irradiation. While irradiation can reduce or eliminate spoilage bacteria or pathogenic microorganisms which may be present in a spoiled food, it cannot improve its sensory properties – the bad appearance, taste or smell will remain.

Treatments such as heat pasteurization, chemical fumigation, and irradiation, however, are effective in destroying or suppressing microbial contamination of food. Heat pasteurization and fumigation have been effectively used in this way for decades to "clean up" foods, specifically to destroy pathogenic microorganisms in milk and other liquid products, and to destroy spoilage microflora or microorganisms and insects in spices and dry foods. These treatments are done intentionally for public health reasons; for example, to destroy microorganisms such as Salmonella, Shigella, and Campylobacter that are associated with food-borne diseases. Irradiation is especially effective as a control measure for pathogenic microorganisms transmitted through solid food, especially foods of animal origin even when in the frozen state.

Food processes such as heating, freezing, chemical treatment, and irradiation are not intended to serve as substitutes for good hygienic practice. Both at the national and international levels, good manufacturing practices (GMPs) govern the handling of specific foods and food products. They must be followed in the preparation of food, whether the food is intended for further processing by irradiation or any other means. An additional step in the further processing of food, such as irradiation, requires a stricter adherence to GMP so that the products reach the final stages at the highest possible quality level.

A similar question was asked of pasteurization when it was first proposed as a means of improving the safety of milk. Pasteurized milk was demonstrated to be safe, practical and fit for the needs of most urban consumers. It was very similar in taste and colour to fresh milk and required no change in consumption or cooking habits. However, the pasteurization of milk did not become a commercial reality for many years after its introduction in the early 1900s. A similar situation has arisen with irradiated food. Although the safety and benefits of food irradiation have been thoroughly documented, the commercial application of the process has been hindered due to some misconception by the general public on its safety and the conservative position of the food industry.

Other processes such as chemical and heat treatments can also kill insects, moulds and microorganisms, including pathogens in food. However, chemicals can leave residues, and heating food, such as canning, changes its texture, colour and flavour and converts it into a cooked product. Irradiation, on the other hand, achieves its effects without significantly raising the temperature of the food, leaving it closer to the unprocessed state. Unlike the fumigants used for disinfestation and quarantine purposes, for example ethylene oxide and methyl bromide, irradiation does not leave residues in the food and is safer to use. Irradiation is unique, however, in its ability to inactivate pathogenic microorganisms, such as Salmonella, E. coli O157:H7 and Campylobacter, in food in the frozen state, particularly in food of animal origin.

Food irradiation has an important role to play in the production of safe, wholesome food just as heat-pasteurization has. At a time when the number of food poisoning outbreaks is on the increase, when fumigants are being phased out, and the consumer is looking for safer, higher quality foods, the overwhelming benefits of food irradiation cannot be overlooked. Irradiation helps to ensure a safer and more plentiful food supply by extending shelf-life and controlling pests and pathogens in food. Most importantly, it is a safe process.

Food Irradiation Facilities

Industrial food irradiation facilities must be licensed, regulated and inspected by national radiological safety and health authorities, many of whom base their rules upon irradiation standards and codes of practice jointly established by the IAEA, FAO and WHO. The common features of all commercial irradiation facilities are the irradiation room and a system to transport the food into and out of the room. The major structural difference between this type of plant and any other industrial building is the concrete shielding (1.5 – 1.8 metres thick) surrounding the irradiation room, which ensures that ionising radiation does not escape to the outside of the room.

In the case of a gamma irradiator, the radionuclide source continuously emits radiation and when not being used to treat food must be stored in a water pool (usually 6 metres in depth). Known as one of the best shields against radiation energy, water absorbs the radiation energy and protects workers from exposure if they must enter the room. In contrast to gamma irradiators, machines producing high-energy electrons operate on electricity and can be switched off.

The transport system employed in a large food irradiation facility is similar to that used for sterilization of medical products and can be either a conveyor or a rail system. In a gamma irradiator, the size of the containers in which the food is moved through the irradiation chamber can vary and pallets up to 1 m^3 may be used. On the other hand, with machines, the bulk or thickness of a product which can be treated is much less and hence there is a fundamental design difference between the two types of irradiator.

Over the past 30 years, laws and regulations have been promulgated to govern operations at industrial irradiators used to process non-food products, such as medical supplies. About 170 such irradiators are operating around the world. The plants, which must be approved by governmental authorities before construction, are subject to regular inspections, audits, and other reviews to ensure that they are safely and properly operated. These types of governmental controls would also be valid for irradiation facilities processing food. For example, the principle of lot traceability is an essential part of process controls, whether the product is a pharmaceutical or a fruit, and irrespective of the technology involved.

At the international level, provisional guidelines for good manufacturing practices (GMPs) and good irradiation practices for a number of foods have been issued by the International Consultative Group on Food Irradiation (ICGFI). They cover all aspects of treatment, handling, and distribution. These guidelines provide a good basis for preparing the detailed protocols needed to implement irradiation on a commercial scale. Some of the guidelines have converted to standards of the American Society for Testing and Materials (ASTM).

The guidelines emphasise that, as with all food technologies, effective quality control systems need to be installed and adequately monitored at critical control points at the irradiation facility. Foods should be handled, stored, and transported according to good manufacturing practices (GMPs) before, during, and after irradiation. Only high-quality food should be accepted for irradiation.

The Codex Alimentarius Commission of FAO and WHO adopted in 1983 a Codex General Standard for Irradiated Foods, and an associated International Code of Practice for the Operation of Radiation Facilities Used for the Treatment of Foods. These standards state that irradiated foods should be accompanied by shipping documents identifying the irradiator, date of treatment, lot identification, dose, and other details of treatment. ICGFI additionally has established an international registry of irradiators that meet standards for good operations. It also organizes training courses for irradiator operators, plant managers, and supervisors on proper processing, with emphasis on GMPs, dosimetry, record-keeping, and lot identification, and for food control officials on proper inspection procedures required for food irradiation processing and trade in irradiated foods.

Any industrial activity includes certain risks to human beings and the environment. One of the risks at irradiation facilities is associated with the potential hazard of accidental exposure to ionizing radiation. Irradiators are designed with several levels of redundant protection to detect equipment malfunction and to protect personnel from accidental radiation exposure. Under normal operating conditions, all exposures of workers to radiation are prevented because the radiation source is shielded. Potentially hazardous areas are monitored and a system of interlocks prevents unauthorized entry into the radiation room while products are being irradiated. Worker safety further rests upon strict operating procedures and proper training. All radiation plants must be licensed. In most countries, regulations require periodic inspection of facilities to ensure compliance with the terms of operating licenses. In the United Kingdom, the Health and Safety Executive has reported to a parliamentary committee that personnel working in the country's 10 irradiation facilities face no unusual dangers: "...the risk is kept under effective control by the use of sophisticated safety control systems. The plants are constructed with very heavy radiation shielding and thus the process presents no risk to the general public. We do not expect that the legalisation of foodstuffs irradiation will present any novel health and safety issues within our area of interest".

Over the past 30 years, there have been a few major accidents at industrial irradiation facilities that caused injury or death to workers because of accidental exposure to a lethal dose of radiation. All of the accidents happened because safety systems had been deliberately bypassed and proper control procedures had not been followed. None of these accidents endangered public health and environmental safety.

In most cases, reports of "accidents" have actually turned out to be operational incidents. Such incidents have caused the irradiator to be shut down but they did not harm anyone or pose a risk to the environment. The distinction between accidents and incidents is used by authorities responsible for safety in all industries. This is the case for many other food technologies, such as canning, fumigation and the agro-chemical industry, which are also potentially hazardous to workers. At irradiation facilities, controls and formal protocols are strictly required to prevent accidents.

The radiation processing industry is considered to have a very good safety record. Today there are about 170 industrial gamma irradiation facilities operating worldwide, a number of which process food in addition to other types of products. Most irradiation facilities are used for sterilizing disposable medical and pharmaceutical supplies, and for processing other non-food items. Facilities are constructed to standard designs with multiple safeguards to protect worker health and safeguard the community should a natural disaster such as an earthquake or tornado occur.

Radioactive materials required for irradiators is transported in lead-shielded steel casks. These casks meet national and international standards modelled upon the Regulations for Safe Transport of Radioactive Materials of the International Atomic Energy Agency (IAEA) and are designed to withstand the most severe accidents, including collisions, punctures, and exposure to fire and water depths. Large quantities of radioactive material are safely shipped all over the world to supply some 170 irradiators processing a variety of goods, mainly medical products such as syringes, physician gloves, sutures, and hospital gowns. From 1955 to date, Canada has shipped approximately 480 million curies of cobalt-60 without any radiation hazard to the environment or release of radioactive materials. Over the same period, approximately one million shipments of radioisotopes for industrial, hospital, and research use were made in North America without radiation accidents. This excellent safety record far exceeds that of other industries shipping hazardous materials such as toxic chemicals, crude oil, or gasoline. The same procedures used so successfully and safely to transport radioactive materials to existing irradiators will of course be used for transporting radioactive materials to any additional irradiators constructed for food processing.

It is impossible for a "meltdown" to occur in a gamma irradiator or for the radiation source to explode. The source of radiation energy used at irradiators cannot produce neutrons which can make materials radioactive, so no nuclear "chain reaction" can occur at an irradiator. The walls of the irradiation cell through which the food passes, the machinery inside the cell, and the product being processed cannot become radioactive. No radioactivity is released into the environment.

It is a misconception that the existence of gamma irradiation facilities will lead to a growing accumulation of radioactive waste material. At gamma irradiators, radionuclide sources, typically cobalt-60 or cesium-137, are used as the sources of radiation energy. These elements decay over time to non-radioactive nickel and non-radioactive barium, respectively. The sources are removed from the irradiator when the radioactivity falls to a low level, usually between 6% and 12% of the initial level (this takes 16 to 21 years for cobalt-60). The elements are then returned in a shipping container to the supplier who has the option of reactivating them in a nuclear reactor or storing them. It has been estimated that when the useful life of the cobalt-60 is finally over, all the used cobalt-60 produced in North America could be stored in a space of about 1.25 cubic metres, which is roughly equivalent to the size of a small office desk.

Basically, the same procedures are followed when an irradiation plant closes down. The sources can be acquired by another user or returned to the supplier, the machinery dismantled, and the building used for other purposes. There is no radiation hazard for the new occupants or the general public.

Safety of Irradiated Food

Irradiation does not make food radioactive. Everything in our environment, including food, contains trace amounts of radioactivity. This means that this trace amount (about 150 to 200 becquerels/kg) of natural radioactivity from elements such as potassium is unavoidable in our daily diets. In countries where food irradiation is permitted, both the sources of radiation and their energy levels I are regulated and controlled. The irradiation process involves passing the food through a radiation field at a set speed to control the amount of energy or dose absorbed by the food. The food itself never comes into direct contact with the radiation source. The maximum allowable energies for electrons and X-rays from the two machine-generated sources of radiation that can be used, are 10 million electron volts (MeV) and 5 MeV, respectively. Even when foods are exposed to very high doses of radiation from these sources, the maximum level of radioactivity would be just one-thousandth of a becquerel per kilogram of food. This is 200,000 times smaller than the level of radioactivity naturally present in food. Food undergoing irradiation does not become radioactive any more than luggage passing through an airport X-ray scanner or teeth that have been X-rayed.

Irradiated foods are those that have been deliberately processed with certain types of radiation energy to bring about some desirable properties (for example, to inhibit sprouting or to destroy food-poisoning bacteria). Apart from foodstuffs, many other materials are commercially irradiated during manufacturing. These include cosmetics, wine bottle corks, hospital supplies and medical products, and some types of food packaging. Radioactive foods, on the other hand, are those that have become accidentally contaminated by radioactive substances from weapons testing or nuclear reactor accidents. This type of contamination is totally unrelated to irradiated food which has been processed for preservation and other purposes.

Since the late 1940s irradiated foods were considered to require careful toxicological investigation before this process could be applied to food manufacturing. In actual fact it was firmly concluded by a study conducted in Germany as far back as 1926 that irradiation did not produce any toxic factors in animal diets. The standard procedure for this purpose was to feed the foodstuff to be tested to laboratory animals and look for possible effects of longevity, reproductive capacity, tumour incidence, and other indicators of the animals' health status.

Several hundred toxicological studies have been conducted on experimental animals over the past four decades. Many animal feeding tests including genetic studies of different types of irradiated food were carried out in many countries including China, Germany, India, Japan, Thailand, the United Kingdom and the USA in the past five decades. FAO, IAEA and WHO convened a number of Joint Expert Committees on the Wholesomeness of Irradiated Foods in 1964, 1969, 1976 and 1980 as data became available to evaluate the safey for consumption of irradiated foods. These evaluations together with those carried out independently by national expert groups in Denmark, France, the Netherlands, Japan, the United Kingdom and the USA demonstrated no toxic effects as a result of consuming irradiated food. Another expert committee evaluated for the WHO in 1992 all literature and data which had been available since 1980; as a consequence, the previous findings were reconfirmed. During September 1997 a study group meeting was organised jointly by the WHO, FAO and IAEA to evaluate the wholesomeness of food treated by high dose irradiation. This group of experts concluded that doses greater than 10 kGy "will not lead to changes in the composition of the food that, from a toxicological point of view, would have an adverse effect on human health".

Among the many extensive animal feeding studies of irradiated food, those conducted at the Raltech Laboratory, USA, are generally acknowledged to be among the best and most statistically powerful of all. The studies involved using chicken irradiated either by a cobalt-60 source or electron machine up to a dose of 58 kGy. Some 134 tonnes of chicken meat, or nearly a quarter of a million birds, were used in the study to compare high-dose irradiation with heat-sterilization of chicken. The study involved chronic feeding studies in mice and dogs, teratology studies and muta-genicity tests. The comprehensive results were reviewed by scientists of the United States Food and Drug Administration (FDA) at the time a petition for low dose irradiation of chicken was submitted in the mid-1980s. No adverse effects from consuming chicken processed with high doses of radiation were reported. The lack of treatment-related effects in the many well-conducted studies provides additional assurance that the consumption of irradiated food does not pose a hazard.

Other types of extensive feeding tests also have been done. Over the last 20 years millions of mice, rats, and other laboratory animals have been bred and reared exclusively on an irradiated diet. The diet, treated at doses between 25 and 50 kGy, has been fed to laboratory animals at many institutions involved in food, drug, and pharmaceutical research in Austria, Australia, Canada, France, Germany, Japan, Switzerland, the United Kingdom, and the USA. No trans-mittable genetic defects – teratogenic or oncogenic – have been observed which could be attributed to the consumption of irradiated diets.

The issue of abnormal chromosomes as a result of eating irradiated food has been more sensationalised than any other. The claims focus on the incidence of "poly-ploidy", which is alleged to result from consumption of products made from wheat immediately after irradiation. Polyploidy means the occurrence of cells containing twice or more the number of chromosomes. Human cells normally have 46 chromosomes. If they are polyploid they could have 92 or even 138 chromosomes. The incidence of polyploid cells is naturally occurring and varies among individuals, and even in one individual from day to day. It can also vary from organ to organ within one individual. The biological significance of polyploidy is unknown. When undertaking studies on polyploidy, it is important that many thousands of cells are counted in order to see the effect of a treatment. As polyploid cells are rare, it is essential that enough cells are observed before any valid conclusions can be reached. It can also be extremely difficult to recognise polyploid cells; if normal (diploid) cells happen to be superimposed on the microscope slide they look very much like one polyploid cell.

Media reports have frequently cited results published in the mid-1970s by a group of scientists from the National Institute of Nutrition (NIN) in Hyderabad, India. This group of scientists reported increases in the frequency of polyploid cells in malnourished children, rats, mice and monkeys that they attributed to consumption of products made from wheat immediately after irradiation at 0.75 kGy. When the report is examined more closely, among other shortcomings, it is found that only 100 cells from each of the five children in each group were counted – an incredibly small sample upon which to base any conclusion. In addition, although the results in each group were averaged, there is no indication of the actual incidence in each child. No polyploidy at all was seen when wheat was irradiated and stored for 12 weeks before consumption.

A number of institutions in India and elsewhere have tried to reproduce the results found at NIN based on information made available to them. Some used absorbed radiation doses as high as 45 kGy. For example, a rat feeding study carried out at the Bhabha Atomic Research Centre (BARC) in Mumbai, with freshly irradiated wheat, in which the incidence of polyploidy was determined by counting 3,000 cells from each animal, showed no effect from consuming the irradiated wheat.

None of the studies carried out came up with results similar to those found at NIN. In order to investigate the reasons for the discrepancy between the results reported by the researchers in Hyderabad and Mumbai a committee of experts was appointed by the Indian government in October 1975 to review the findings. In 1976, the report of the Committee was very critical of the work of the Hyderbad authors and concluded that the available data failed to demonstrate any mutagenic potential of irradiated wheat. A number of national scientific committees and independent researchers in Australia, Canada, Denmark, France, the United Kingdom, and the USA also have evaluated the alleged incidence of polyploidy. They all concluded that the study was simply unacceptable and the reported data from NIN do not support the incidence of increased polyploidy.

In the early 1980s, eight feeding studies using several irradiated food items, including irradiated wheat, were conducted in China using human volunteers. More than 400 individuals consumed irradiated food under controlled conditions for 7 to 15 weeks. One focus of the research was the possibility of chromosomal changes. Seven of the eight experiments involved investigation of chromosomal aberrations in 382 individuals. No significant difference between the number of chromosomal aberrations in the control and test groups were discovered in any of the experiments. Incidence of polyploidy in those who consumed non-irradiated food and those who consumed irradiated samples were within normal range of the overall value of polyploid cells in participants.

Although not aimed at testing the safety of irradiated foods it is worth noting at this point that radiation-sterilized foods are used in the diet of severely ill patients. A number of hospitals in the USA and the United Kingdom used irradiated foods for patients who have to be kept in a completely sterile environment due to their susceptibility to bacterial or viral infections. Patients undergoing chemotherapy, or organ transplant patients who receive immunosuppressive medication may be fed only sterilized foods for weeks or even months. Supplementing heat-sterilized foods with radiation-sterilized items can provide more varied, more palatable, and more nutritious menus for these patients. Irradiated foods were fed to such patients at the Fred Hutchinson Cancer Research Center, Seattle, USA for several years during the mid 1970s with excellent results.

The so-called "radiolytic" products produced in irradiated food have proven to be familiar ones, such as glucose, formic acid, acetaldehyde, and carbon dioxide, that are naturally present in foods or are formed by thermal processing (thermo-lytic products). The safety of these radiolytic products has been examined very critically, and no evidence of their harmfulness has been found.

The United States Food and Drug Administration (FDA) has estimated that the total amount of undetected radiolytic products that might be formed when food is irradiated at a dose of 1 kGy would be less than 3 milligrams per kilogram of food or less than 3 parts per million.

The fact that irradiation causes the formation of free radicals – which in scientific terms are atoms or molecules with an unpaired electron – and that these are quite stable in dry foods has often been mentioned as a reason for special caution with irradiated dry foods. However, free radicals are also formed by other food treatments, such as toasting of bread, frying, and freeze drying, and during normal oxidation processes in food. They are generally very reactive, unstable structures, that continuously react with substances to form stable products. Free radicals disappear by reacting with each other in the presence of liquids, such as saliva in the mouth. Consequently, their ingestion does not create any toxicological or other harmful effects.

This has been confirmed by a long-term feeding study carried out a the Federal Research Centre for Nutrition in Karlsruhe, Germany. This study was especially designed to look for possible effects of a diet containing a high free radical concentration. Animals were fed a very dry milk powder irradiated with electrons at 45 kGy. No mutagenic effects were noted and no tumours were formed. Nine generations of rats were continually fed this diet without any indication of toxic effects. Similarly, a slice of toasted bread (non-irradiated), which actually contains more free radicals than very dry foods that have been irradiated, can be expected to be harmless.

Under Good Manufacturing Practices (GMPs), irradiating food of animal origin to ensure its hygienic quality does not increase the risk from botulism any more than other "sub-sterilizing" food processes, such as pasteurization. It is true that bacterial spores such as those of Clostridium botulinum are resistant to most preservation treatments, including low doses of irradiation. However, these spores are usually present in relatively low numbers and, although they survive sub-sterilizing doses of irradiation, other microorganisms also survive irradiation can grow, cause spoilage and inhibit the growth of Clostridium botulinum. The survival of spores is therefore not considered to introduce any additional hazard in irradiated foods than in food subjected to other sub-sterilizing heat-treatments, for example, in pasteurized or cooked foods.

Food treated by irradiation or traditional pasteurization must be handled, packaged, and stored following good manufacturing practices (GMPs). Doing so prevents the growth and toxin production of Clostridium botulinum. Alternatively, high-dose irradiation (30-60 kGy) can be used to destroy any Clostridium botulinum spores present in the food.

Some types of clostridia cause more concern than others. Clostridium botulinum Type E, for example, is found at low levels in fish and seafood caught in some areas. It can grow and produce toxin even when the food is refrigerated at temperatures as low as 4° C. Thus, fish and seafood, including their products, treated by any of the sub-sterilizing processes, including irradiation, must be kept at 3° C or below at all times during marketing. Most other types of Clostridium botulinum cannot grow and produce toxin at temperatures below 10° C. Good manufacturing practices (GMPs) require that raw foods such as fish, meat, and chicken are stored at a specific temperature, whether irradiated or not, to prevent the growth of Clostridium botulinum. It has been concluded by various workers in this field of research that low-dose irradiation does not increase the risk from sporogenous bacteria. On the contrary, it has been emphasised that low-dose irradiation can increase the safety of foods.

Only foods of good hygienic quality should be irradiated. In this respect, irradiation does not differ from heat pasteurization, freezing, or other food processes. While these processes can destroy bacteria, they may not destroy preformed toxins and viruses already in the food. It is very important that foods intended for processing – by whatever method – are of good quality and handled and prepared according to GMPs established by national or international authorities. In some cases, strict regulations prohibit distribution of some foods. Many countries, for example, do not permit oysters to be harvested from areas known to be contaminated with raw sewage because of the danger of hepatitis viruses. No food processing methods should be used to substitute for GMPs in food production and handling.

Nutritional Quality of Irradiated Foods

As irradiation is a 'cold process', that is, it does not substantially raise the temperature of the food being processed, nutrient losses are small and often significantly less than losses associated with other methods of preservation such as canning, drying and heat pasteurization. Much of the early work on irradiation examined foods treated at sterilizing doses, but since recent applications often use doses well below 10 kGy, a realistic evaluation of the nutritional adequacy of irradiated food should be based on results of experiments carried out using doses likely to be used in commercial practice. The change in nutritional value caused by irradiation depends on a number of factors. These include the irradiation dose to which the food has been exposed, the type of food, packaging, and processing conditions, such astemperature during irradiation and storage time.

Carbohydrates, proteins and fats are the main components of foods. These macronutrients provide energy and serve as building blocks for the growth and maintenance of the body. Extensive research has shown that carbohydrates, proteins, and fats, undergo little change during irradiation even at doses over 10 kGy. Similarly, the essential amino acids, minerals, trace elements and most vitamins do not suffer significant losses.

Different types of vitamins have varied sensitivity to irradiation and to some other food processing methods. The sensitivity of the vitamins to irradiation depends on the complexity of the food system and the solubility of the vitamins in water or fat. Irradiation of vitamins in pure solution results in considerable destruction of these compounds thus some reports in literature have overestimated the losses. For example, vitamin B1 (thiamin) in aqueous solution showed 50% loss after irradiation at 0.5 kGy, while irradiation of dried whole egg at that dose caused less than 5% destruction of the same vitamin. This is due to the mutually protective action of various food constituents on each other. Vitamin losses can be minimized by irradiating the food in frozen form or by packaging it in an inert atmosphere such as under nitrogen.

Four vitamins are recognised as being highly sensitive to irradiation: B1, C (ascorbic acid), A (retinol) and E (a-tocopherol). However, B1 is even more sensitive to heat than to irradiation. It has been demonstrated that pork and beef sterilized by irradiation retain much more vitamin B1 than canned meat sterilized thermally.

Seemingly conflicting results of low versus high losses of vitamin C for some irradiated foods may be attributed to differences in analytical approaches used by researchers. Some have measured only ascorbic acid, while others have measured total ascorbic acid, a mixture of ascorbic acid and dehydroascorbic acid. Both acids have vitamin C biological activity and are easily transformed from one to the other. If only ascorbic acid were measured, any apparent reduction in vitamin C level would be exaggerated. Research has shown that the natural differences in total vitamin C content of four varieties of strawberry are much greater than the reduction which occurs on irradiation. With, for example, potatoes it has been demonstrated that although irradiation does reduce vitamin C content, cooking and storage also have a significant effect. The benefit of irradiating potatoes is to inhibit sprouting during storage. Following six months of storage the vitamin C content of irradiated and unirradiated potatoes have been shown to be similar. Since the optimal dose for irradiation treatment of fruit and vegetables, is generally below 2 kGy, effects on vitamin C at higher doses are irrelevant.

The significance of any losses of vitamins E and A due to irradiation are marginal because the main sources of these vitamins in the human diet are butter and milk and these are unsuitable for irradiation treatment. Irradiation has practically no effect on the levels of beta carotene and other carotenoids, the precursors of vitamin A, formed in fruits during ripening.

On the whole, the effects of irradiation on the nutritional value of foods are minimal and these observations are substantiated by the results of many feeding studies which have been undertaken to establish the wholesomeness of irradiated food. It should also be remembered that irradiated food will be consumed as part of a mixed diet, and therefore the process will have little impact on the total intake of specific nutrients.

The Joint Expert Committee of the Food and Agriculture Organization (FAO), World Health Organization (WHO), and International Atomic Energy Agency (IAEA), which examined these and other issues, stated in its conclusions in 1980 that irradiation does not introduce special nutritional problems in food. This was also the finding of the group of experts who convened at a meeting organised by the FAO, IAEA and WHO in Geneva, Switzerland in 1997 to discuss the effects of high dose irradiation. It was concluded at this meeting that doses greater than 10 kGy "will not lead to nutrient losses to an extent that would have an adverse effect on the nutritional status of individuals or populations".

Packaging of Irradiated Foods

With the exception of such applications as sprout inhibition in potatoes or onions, insect disinfestation in bulk grains, or delay of post-harvest ripening of fruits, irradiation of foodstuffs is usually carried out on packaged food items. There may be different reasons for this: prevention of microbial reinfection or insect exposure, prevention of water loss, exclusion of oxygen, prevention of mechanical damage during transport, or simply improved handling and marketing. The packaging material used must not release radiation-induced reaction products or additives onto the food, nor should it lose functional qualities such as mechanical strength, seal stability, or imper meability to water upon irradiation.

Results of extensive research have shown that almost all commonly used food packaging materials tested are suitable for use at any dose likely to be applied to food including sterilization treatment. Only packaging materials which have been specifically authorized for such use may be subjected to irradiation of prepackaged foods.

Various types of packaging materials have been approved for use when food is irradiated. Their suitability for food intended for irradiation has been studied in Canada, the United Kingdom, the USA, and a few other countries. A number of food packaging materials were approved for use in food irradiation by the United States Food and Drug Administration (FDA) more than 20 years ago. More recently, Canada, India and Poland have approved additional materials, including a multi-layered polyethylene film, as safe for packaging foods to be irradiated.

Sophisticated tests have been used to evaluate the effect of radiation on plastic and other types of packaging materials. Researchers looked at the post-irradiation stability, mechanical strength, and permeability to water and gases of the packaging materials, and at the extract-ability of the plastics, additives, and adhesives.

Plastic films laminated with aluminium foil are routinely sterilized by radiation as part of the manufacturing process. They are used for hermetically sealed "bag-in-a-box" products, such as tomato paste, fruit juices, and wines.

Other aseptic packaging materials, dairy product packaging, single-serving containers (for example, for cream), and wine bottle corks are also routinely sterilized by irradiation prior to filling and sealing to prevent product contamination.

Other types of materials used to wrap food or other products also are routinely processed by radiation in many countries. The radiation process is used to "cross-link" the material's polymer chains for greater strength and heat resistance, and for producing plastics with special properties (for example, heat-shrink wrap).

Food Irradiation Costs

Any food process will add cost. In most cases, however, food prices do not necessarily rise just because a product has been treated. Many variables affect food costs, and one of them is the cost of processing. Canning, freezing, pasteurization, refrigeration, fumigation, and irradiation add cost to the product. These treatments will also bring benefits to consumers in terms of availability and quantity, storage life, convenience, and improved hygiene of the food. Reduced losses will bring revenue to producers and traders, thus in turn, compensating treatment costs.

The major factors influencing the economics of food irradiation using cobalt-60 include: irradiation design parameters such as applied dose, packing density of the products, handling conditions (dry versus perishable products), dose uniformity and throughput; capital costs consisting of the irradiator, radiation source, spare parts for linear accelerators, warehouse capacity; and operating costs such as salaries, utilities, replenishments of cobalt-60, maintenance, etc.

Irradiation costs range from US $10 to $15 per tonne for a low-dose application to US $100 to $250 per tonne for a high-dose application (for example, to ensure hygienic quality of spices). These costs are competitive with alternative treatments. In some cases, irradiation can be considerably less expensive. For disinfestation of fruit to satisfy the quarantine requirements of an importing country, for example, it has been estimated that the cost of irradiation would be only 10% to 20% of the cost of vapour-heat treatment.

Electron beam irradiators may have economic advantages over gamma irradiators where product throughput is large, the particle size or thickness of the product being treated is small, and where continuous treatment is possible by integrating the irradiator into the production line. As a result, they may be more efficient than gamma irradiators for treating large volumes of domestic or imported grains. In addition, these machine-type irradiators, based on electron acceleration rather than radionuclides, may not require as extensive regulatory approvals.

The cost to build a commercial cobalt-60 food irradiation plant is in the range of US $3 million to $5 million, depending on its size, processing capacity, and other factors. This is within the range of plant costs for other technologies. For example, a moderately-sized, ultra-high temperature plant for sterilizing milk, fruit juices, and other liquids costs about US $2 million. A small vapour-heat treatment plant for disinfestation of fruits costs about US $1 million.

Often the capital costs of irradiation equipment are seen as prohibitive, even though low operating costs for most commodities make per unit costs very competitive with other treatments. Commercial contract multipurpose-irradiators operate in many countries offering irradiation services at reasonable cost. Most of these facilities successfully combine irradiation of various food products and treatment of other non-food items such as cosmetics, pharmaceutical and disposable medical products. Since irradiation gives the added economic benefit of prolonged fresh market life for many foods, decreased waste and increased market potential of the food should be considered in a cost-benefit analysis.

Trade in Irradiated Foods

Food imports and exports are important to the health and economy of nations and people, yet trade barriers caused by pests, diseases and food safety issues continually threaten or inhibit trade. Several technologies work to remove trade barriers. Irradiation is one such technology that could assist in the improvement of trade.

Some irradiated foods such as spices and dried vegetable seasonings, as well as food ingredients such as mechanically deboned poultry meat, have entered international commerce for use mainly by the food industry in various types of processed food. The nature of the spice trade requires that spices, for example pepper from various sources, be mixed to achieve certain grades to satisfy market demand. Thus, it is possible that only a portion of the spices within one single shipment has been irradiated. The production of, and trade in, irradiated spices have increased significantly in recent years from about 5,000 tonnes in 1990 to over 30,000 tonnes in 1994 to over 60,000 tonnes in 1997. Approximately 30,000 tonnes of irradiated spices and dried vegetable seasonings were produced in the USA alone in 1997.

Fresh fruits and vegetables could be irradiated to overcome quarantine barriers against fruit flies in the near future. The United States Department of Agriculture (USDA) has accepted irradiation as a quarantine treatment against major fruit fly species regardless of host. A policy to this effect was issued by the USDA in 1996. Phytosanitary guidelines for the use of irradiation as quarantine treatment will likely be applied in other countries. South East Asia, for example, is in the process of implementing a harmonized protocol on the use of irradiation as a quarantine treatment of horticultural commodities.

Labelling

Some national regulations require that irradiated food be labelled with a statement indicating the treatment and, often, with an international logo known as the radura symbol. Experience with market trials and commercial sale of irradiated food has proven that informed consumers are not against irradiated food but prefer it to be labelled as such. The label provides consumers with the opportunity to choose. Label statements can also be used to state why products are irradiated. It has been demonstrated that people are more likely to buy irradiated food labelled with a statement conveying the positive benefits of the technology, for example, 'Irradiated to control microbes' or 'Irradiated to retard spoilage'.

For irradiated foods that are not packaged, such as bulk containers of fruit and vegetables, retailers in some countries are required to display the logo and phrase. Labelling regulations do, however, differ between countries. For example, in the USA, labelling requirements apply only to whole foods that have been irradiated but not to irradiated ingredients in a food. In the European Union it is proposed that food containing irradiated ingredients such as spices, but which are not themselves irradiated, must be labelled regardless of the percentage of irradiated product which has been incorporated.

Many governments have introduced regulations requiring labelling of irradiated food but not to those treated by competitive treatments such as fumigation. However, in August 1998 the United States Food and Drug Administration (FDA) amended the labelling requirements for irradiated food – a statement disclosing irradiation treatment is not required to be any more prominent than the declaration of ingredients.

A government can deny entry of any product into its territory. However, under the provision of the Agreement on the Application of Sanitary and Phytosanitary Measures (SPS), being enforced by the World Trade Organization (WTO), such a government (if a member of WTO) may be requested to furnish scientifically-based justification for regulations that are stricter than the only recognized international standards for food, which are the guidelines, and recommendations of the Codex Alimentarius Commission (food safety), the International Plant Protection Convention (IPPC) (plant protection and quarantine), and the International Office of Epizootics (animal health and quarantine). With the existence of the Codex General Standard for Irradiated Foods, which recognizes the safety and effectiveness of food irradiation, and the endorsement of irradiation as a quarantine treatment of fresh agricultural produce by regional plant protection organizations operating within IPPC, irradiated food treated according to the principle of the Codex Standard can no longer be denied entry into countries on scientific grounds.

Detection Methods for Irradiated Food

The need for reliable and routine tests to determine whether or not food has been irradiated arose as a result of the progress made in commercialisation of the food irradiation process, greater international trade in irradiated foods, differing regulations relating to the use of the technology in many countries, and consumer demand for clear labelling of the treated food. Although not essential for management of the process, it was envisaged that the availability of such tests would help strengthen national regulations on irradiation of specific foods, and enhance consumer confidence in such regulations. The availability of reliable identification methods would be of assistance in establishing a system of legislative control, and help to achieve acceptance of irradiated foods by consumers. In fact, an International Conference on Acceptance, Control of, and Trade in Irradiated Food held in Geneva in 1988 recommended that "governments should encourage research into methods of detection of irradiated foods so that administrative control of irradiated food once it leaves the facility can be supplemented by an additional means of enforcement, thus facilitating international trade and reinforcing consumer confidence in the overall process".

Since the mid 1980s extensive research has resulted in the development of a range of tests which can be used to reliably determine the irradiation status of a wide variety of food. The methods which have been studied most extensively and which have the greatest scope of application include electron spin resonance (ESR) spectroscopy, thermoluminescence (TL), and monitoring the formation of long-chain hydrocarbons and 2-alkylcyclobutanones. These methods have been successfully evaluated in a number of interlaboratory blind trials with the result that, in 1996, five tests were adopted as standard reference methods for the detection of irradiated food by the European Committee for Normalisation (CEN). These in turn are being adopted by some national authorities, such as Germany and the United Kingdom. More tests are being considered by CEN for implementation as reference methods.

Consumers of Irradiated Foods

The ultimate test for any product or process is the market place, as it is the consumer who determines whether or not a product is better than previous or competitive products. Such is the case for irradiated food which, at the end of the day, will not be a technical or marketing success unless it is accepted by the consumer. Progress in the commercial use of irradiation has been slow, mainly because of misunderstanding. Many people mistakenly fear, for instance, that the process may induce radioactivity in the food product and that irradiation will result in the formation of toxic by-products in food. Given these fears, consumers often find it difficult to evaluate the benefits of this processing technique objectively.

While many consumers are unfamiliar with food irradiation, consumer research shows that, as more and more factual information is provided, the public increasingly views irradiation in a more positive light. In fact, some studies have even shown a consumer willingness to pay a premium price for irradiated products. Consumers indicate that endorsements by a respected health authority increase their confidence in the safety of this technology. A United States Department of Agriculture (USDA) funded project in California and Indiana evaluated the impact of a brief educational programme on community leaders' attitudes to and knowledge of food irradiation. After viewing a 10-minute video on food irradiation, those likely to try irradiated food increased from 57% to 83%.

An extensive marketing and educational programme was conducted in South Africa prior to the introduction of irradiated foods in the market. With regard to irradiated shelf-stable meat products, a marketing survey among the general population found that initially 15% of people surveyed indicated they were likely to purchase the irradiated food. After receiving visual information, those willing to buy increased to 54%. After receiving information and tasting the food, 76% indicated they would purchase the irradiated shelf-stable product, while 5% indicated that they probably would not buy.

Many market tests of irradiated foods have been carried out in several countries during the 1980s and 1990s, andto date, all have been successful. Market tests and the ongoing sales of irradiated foods provide the best source of information on whether irradiated foods meet consumer needs and wants. They provide excellent information about consumer acceptance and facilitate the commercialization of irradiation.

Irradiated mangoes sold well in Florida in 1986, and in March 1987, irradiated Hawaiian papayas outsold their identically priced non-irradiated counterparts by more than ten to one. Irradiated apples marketed in Missouri were also favourably received. In March 1992, a retailer in the Chicago area featured irradiated strawberries, grapefruit and juice oranges. The irradiated produce outsold the non-irradiated produce by ten to one. Irradiated tomatoes, mushrooms and onions were later marketed with similar sales success. In the second year of operation and thereafter, irradiated produce continued to outsell non-irradiated produce by twenty to one. In 1995 the same store and several Midwest markets in the USA sold tropical fruits, including papaya, litchi and starfruit, from Hawaii in conjunction with a study to determine the potential of irradiation as a quarantine treatment. As of the end of August 1998, the total amount of Hawaiian fruits irradiated and sold in the USA has been about 280,00 pounds demonstrating that consumers are willing to buy these fruits, and do so repeatedly. In a small scale test of irradiated poultry in Kansas, USA in 1995, irradiatedpoultry captured 60% of the market share when priced 10% lower than store brand, 39% when priced equally, and 30% when priced 10% higher. In 1996 when another test was conducted, the market share increased to 63% when the irradiated poultry was priced 10% lower than the store brand, 47% when priced equally, and 18% and 17% when priced 10% and 20% higher. The irradiated product sold better in the more up-market store, capturing 73% of the market when priced 10% lower, 58% when priced equally, and 31% and 30% when priced 10% and 20% higher. This is consistent with other attitude surveys and market place data that indicate irradiation is more acceptable in up-scale markets.

Irradiated onions and garlic were first sold in a supermarket in the Buenos Aires area in 1985. Prior to the first marketing, consumers were informed about food irradiation in the local TV, radio and press. Within three days of marketing, the entire 10 tonnes of irradiated product were sold. Subsequent trials gave similar results.

Irradiated dried fish and onions have been successfully test marketed in Bangladesh. In China, numerous irradiated foods have been tested during 1980s and early 1990s including apples, garlic, seasonings, meat products, sweet potato wine, potatoes, tomatoes, and dehydrated vegetables. Successful test markets using brown rice, mungbean, and glutinous rice have been conducted in Indonesia.

Several irradiated foods have been successfully test marketed in Thailand, including irradiated onions, fragrant rice and sweet tamarind. Nham, fermented pork sausage consumed raw in Thailand, is often contaminated with Salmonella and occasionally with Trichinella spiralis. In 1986, labelled irradiated Nham was sold side by side with the traditional product. A consumer survey showed that 34% of the buyers selected irradiated Nham out of curiosity and 66% considered it safer from harmful microorganisms. Satisfaction was high, with 95% of customers indicating that they would purchase irradiated Nham again. During the three-month test, irradiated Nham outsold the non-irradiated product by a ratio of ten to one. This product is presently available in Thailand on a commercial basis.

From 1994 to 1996 several irradiated products including dried mushrooms, dried meat, dried vegetables and dried fish, were market tested in the Republic of Korea and found acceptable to consumers. Irradiated potatoes, onions, and dried fruits were marketed at different times in various shops in Peshawar, Pakistan where consumers found them to be more acceptable than the non-irradiated products. One tonne each of onions and potatoes were test marketed at a provincial fruit and vegetable show in January and February 1991. Thirty-nine per cent of 300 consumers who completed a survey form said they were willing to buy irradiated food and to convince others to buy it, and 57% thought food irradiation should be commercialized. Market tests with irradiated onions have been carried out in the Philippines since 1985 and sales of irradiated onions have always been high. When irradiated onions and potatoes were marketed in two cities in Poland in 1988, 97% of the consumers responding to a survey evaluated the products positively, and say they would like to buy them again.

Irradiated strawberries were test marketed in Lyon, France in May and June 1987. Two tonnes of products, packed in covered plastic trays, labelled 'Protected by Ionization', and priced 30% higher than non-irradiated products, sold well. Market tests have also involved irradiated chicken breast meat and Camembert cheese.

In South Africa between 1978 and 1979 irradiated potatoes, mangoes, papayas, and strawberries were sold in 20 supermarkets where they were judged acceptable by 90% of buyers. Over a six-year period, several popular dishes, for example grilled chicken, curried chicken, bacon, curried beef, and a Malaysian dish called bobotie, were evaluated by a large number of groups, including hikers and sea voyagers. High acceptance was indicated by researchers. Approximately 200 members of the Defence Force tested the products and showed overwhelming preference for the irradiated product over freeze-dried and canned counterparts.

Thus, all of the marketing trials carried out have clearly demonstrated that consumers are receptive towards irradiated food and will select it in preference to a non-irradiated equivalent when they perceive benefits. In all of these trials it was evident that informed consumers will accept irradiated foods.

Irradiated Food Items being Marketed at Retail Level

Several irradiated foods are used directly by the food industry. For example irradiated spices and mechanically-deboned poultry meat are used for manufacturing various types of processed food. Many irradiated foods are also being marketed at the retail level.

Fresh Fruits

Since the first commercial food irradiator in the USA began operating in Mulberry, near Tampa, Florida in 1992, irradiated strawberries, tomatoes and citrus fruits have been marketed at several retail outlets in Florida and

Illinois. The USDA issued its policy in 1996 to accept irradiation as a quarantine treatment of fresh fruits against fruit flies regardless of the host commodities. Fruits from Hawaii including papaya, rambutan, litchis and cherimoya which are natural hosts of fruit flies have been irradiated and marketed under a special permission from the USDA at the retail level in several States in the USA since 1995. All irradiated products are labelled with the irradiation logo and a statement 'treated by irradiation' either on the package or at the point of sale. The USDA policy has also made it possible for irradiation to be used as a quarantine treatment of fresh fruits from other countries against major species of fruit flies, regardless of host commodity as long as the fruit is not a host for other quarantine pests.

In China, irradiated apples have been marketed at the retail level in Shanghai and other cities since the early 1990s.

Spices and Dried Vegetable Seasonings

Irradiated spices and dried vegetable seasonings have been retail marketed in South Africa over the past 10 years and the volume is increasing. In fact, irradiation is used so routinely by the spice trade in South Africa that it would be difficult to find spices treated by some other means (fumigation, heat) in the country. A variety of processed food (for example, sauces, salad dressings, sandwich spread) also incorporate irradiated spices and vegetable seasonings. All irradiated products have to be labelled with an irradiation logo plus the word 'Radurised'. Since 1995, irradiated spices and dried vegetable seasonings have been marketed at retail levels in Belgium. Irradiated spices, condiments and seasonings are also available in China.

Frog Legs

Because of strict microbiological specifications in France, most, if not all, frog legs marketed in the country have been treated by irradiation to ensure their hygienic quality. The product has to be labelled 'treated by ionization' and can be purchased in most French food markets.

Onions, Garlic

Vidalia onions have been irradiated in Florida, to prevent sprouting, and marketed at retail level in Chicago since 1992. Irradiated garlic has been sold in increasing quantities in several cities in China since the early 1990s. During 1995-98, about 166,000 tonnes of garlic were irradiated and marketed across China. All products are labelled to indicate the treatment.

Chicken

Following the approval of the United States Food and Drug Administration (FDA) and the quality control programme for irradiated poultry in 1993, small quantities of irradiated chicken have been offered for sale in some retail outlets in Florida, Illinois, Iowa and Kansas with success.

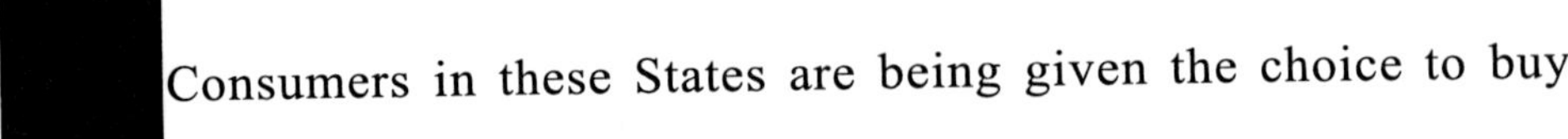

Consumers in these States are being given the choice to buy irradiated chicken without pathogens such as Salmonella for the first time.

Fermented Pork Sausages

Irradiated fermented pork sausages treated for controlling pathogenic microorganisms and parasites, are gaining popularity since their first market trial in 1986. During 1997, about 80 tonnes of Nham were irradiated in Thailand. Increasing quantities of this irradiated product are being supplied to supermarkets in Bangkok. In addition to marketing in Bangkok, the producers have developed new markets in the north and northwest of Thailand where Nham is a staple part of the diet. There is a widespread demand for this irradiated product as the risk from infection by Salmonella and Trichinella spiralis has been removed. The irradiation logo and a statement indicating irradiation treatment are required on the label.

Other Food Products

Since a semi-commercial irradiator in Chittagong, Bangladesh went into operation in 1993, small quantities of irradiated dried fish (for insect control) have been available in the market in Chittagong and other cities in Bangladesh, with labelling indicating irradiation treatment.

Shelf-stable, ready-to-eat meals are commercially available in South Africa. At present, 12 such meals are in existence, including beef curry, beef stroganoff, chicken curry, lasagna and a Malaysian dish called bobotie. These meals are labelled as irradiated and have a shelf-life of greater than two years, making them ideal for outdoor activities such as hiking, camping, yachting, safaris and mountaineering.

Successful market trials of other irradiated foods such as rice, mungbeans, potatoes and onions in several countries in recent years will likely lead to further commercialization of irradiated food in the near future.

The actual sale of irradiated food in the market in several countries has clearly demonstrated that consumers will accept irradiated food if they have the choice.

References

Henkel, J. 1998. Irradiation – A safe measure for safer foods. *FDA Consumer,* May–June, Pages 12-17.

ICGFI. 1998 "Irradiation and Trade in Food and Agriculture Products". *International Consultative Group on Food Irradiation Policy Document,* Vienna.

Satin, M. 1993. *Food Irradiation–A Guidebook.* Technomic Publishing Co., Inc., Lancaster, USA.

Thakur, B.R. and Singh, R.K. 1994. Food irradiation – chemistry and applications. *Food Reviews International*, Volume 10, Part 4, Pages 437-473.

WHO. 1988. *Food Irradiation. A Technique for Preserving and Improving the Safety of Food.* World Health Organization, Geneva.

Wilkinson, V.M. and Gould, G.W. 1996. *Food Irradiation – A Reference Guide.* Butterworth-Heinemann, Oxford, UK.

8

Food Canning

Canning can be a safe and economical way to preserve quality food at home. Disregarding the value of your labor, canning homegrown food may save you half the cost of buying commercially canned food. Canning favorite and special products to be enjoyed by family and friends is a fulfilling experience and a source of pride for many people.

Many vegetables begin losing some of their vitamins when harvested. Nearly half the vitamins may be lost within a few days unless the fresh produce is cooled or preserved. Within 1 to 2 weeks, even refrigerated produce loses half or more of some of its vitamins. The heating process during canning destroys from one-third to one-half of vitamins A and C, thiamin, and ribofavin. Once canned, additional losses of these sensitive vitamins are from 5 to 20 percent each year. The amounts of other vitamins, however, are only slightly lower in canned compared with fresh food. If vegetables are handled properly and canned promptly after harvest, they can be more nutritious than fresh produce sold in local stores.

The advantages of home canning are lost when you start with poor quality fresh foods; when jars fail to seal properly; when food spoils; and when favors, texture, color, and nutrients deteriorate during prolonged storage.

How Canning Preserves Foods?

The high percentage of water in most fresh foods makes them very perishable. They spoil or lose their quality for several reasons:

- growth of undesirable microorganisms—bacteria, molds, and yeasts,
- activity of food enzymes,
- reactions with oxygen,
- moisture loss.

Microorganisms live and multiply quickly on the surfaces of fresh food and on the inside of bruised, insect-damaged, and diseased food. Oxygen and enzymes are present throughout fresh food tissues.

Proper canning practices include:

- carefully selecting and washing fresh food,
- peeling some fresh foods,
- hot packing many foods,
- adding acids (lemon juice or vinegar) to some foods,
- using acceptable jars and self-sealing lids,
- processing jars in a boiling-water or pressure canner for the correct period of time.

Collectively, these practices remove oxygen; destroy enzymes; prevent the growth of undesirable bacteria, yeasts, and molds; and help form a high vacuum in jars. Good vacuums form tight seals which keep liquid in and air and microorganisms out.

Ensuring Safe Canned Foods

Growth of the bacterium Clostridium botulinum in canned food may cause botulism—a deadly form of food poisoning. These bacteria exist either as spores or as vegetative cells. The spores, which are comparable to plant seeds, can survive harmlessly in soil and water for many years. When ideal conditions exist for growth, the spores produce vegetative cells which multiply rapidly and may produce a deadly toxin within 3 to 4 days of growth in an environment consisting of:

- a moist, low-acid food
- a temperature between 40° and 120°F
- less than 2 percent oxygen.

Botulinum spores are on most fresh food surfaces. Because they grow only in the absence of air, they are harmless on fresh foods.

Most bacteria, yeasts, and molds are diffcult to remove from food surfaces. Washing fresh food reduces their numbers only slightly. Peeling root crops, underground stem crops, and tomatoes reduces their numbers greatly. Blanching also helps, but the vital controls are the method of canning and making sure the recommended research-based process times, found in these guides, are used.

The processing times in these guides ensure destruction of the largest expected number of heat-resistant microorganisms in home-canned foods. Properly sterilized canned food will be free of spoilage if lids seal and jars are stored below 95°F. Storing jars at 50° to 70°F enhances retention of quality.

Food Acidity and Processing Methods

Whether food should be processed in a pressure canner or boiling-water canner to control botulinum bacteria depends on the acidity of the food. Acidity may be natural, as in most fruits, or added, as in pickled food. Low-acid canned foods are not acidic enough to prevent the growth of these bacteria. Acid foods contain enough acid to block their growth, or destroy them more rapidly when heated. The term "pH" is a measure of acidity; the lower its value, the more acid the food. The acidity level in foods can be increased by adding lemon juice, citric acid, or vinegar.

Low-acid foods have pH values higher than 4.6. They include red meats, seafood, poultry, milk, and all fresh vegetables except for most tomatoes. Most mixtures of low-acid and acid foods also have pH values above 4.6 unless their recipes include enough lemon juice, citric acid, or vinegar to make them acid foods. Acid foods have a pH of 4.6 or lower. They include fruits, pickles, sauerkraut, jams, jellies, marmalades, and fruit butters.

Although tomatoes usually are considered an acid food, some are now known to have pH values slightly above 4.6. Figs also have pH values slightly above 4.6. Therefore, if they are to be canned as acid foods, these products must be acidifed to a pH of 4.6 or lower with lemon juice or citric acid. Properly acidifed tomatoes and fgs are acid foods and can be safely processed in a boiling-water canner.

Botulinum spores are very hard to destroy at boiling-water temperatures; the higher the canner temperature, the more easily they are destroyed. Therefore, all low-acid foods should be sterilized at temperatures of 240° to 250°F, attainable with pressure canners operated at 10 to 15 PSIG. PSIG means pounds per square inch of pressure as measured by gauge. The more familiar "PSI" designation is used hereafter in this publication. At temperatures of 240° to 250°F, the time needed to destroy bacteria in low-acid canned food ranges from 20 to 100 minutes. The exact time depends on the kind of food being canned, the way it is packed into jars, and the size of jars. The time needed to safely process low-acid foods in a boiling-water canner ranges from 7 to 11 hours; the time needed to process acid foods in boiling water varies from 5 to 85 minutes.

Process Adjustments at High Altitudes

Using the process time for canning food at sea level may result in spoilage if you live at altitudes of 1,000 feet or more. Water boils at lower temperatures as altitude increases. Lower boiling temperatures are less effective for killing bacteria. Increasing the process time or canner pressure compensates for lower boiling temperatures. Therefore, when you use the guides, select the proper processing time or canner pressure for the altitude where you live. If you do not know the altitude, contact your local county Extension agent. An alternative source of information would be the local district conservationist with the Soil Conservation Service.

Equipment and methods not recommended

Open-kettle canning and the processing of freshly flled jars in conventional ovens, microwave ovens, and dishwashers are not recommended, because these practices do not prevent all risks of spoilage. Steam canners are not recommended because processing times for use with current models have not been adequately researched. Because steam canners do not heat foods in the same manner as boiling-water canners, their use with boiling-water process times may result in spoilage. It is not recommended that pressure processes in excess of 15 PSI be applied when using new pressure canning equipment. So-called canning powders are useless as preservatives and do not replace the need for proper heat processing. Jars with wire bails and glass caps make attractive antiques or storage containers for dry food ingredients but are not recommended for use in canning. One-piece zinc porcelain-lined caps are also no longer recommended. Both glass and zinc caps use fat rubber rings for sealing jars, but too often fail to seal properly.

Ensuring High-quality Canned Foods

Begin with good-quality fresh foods suitable for canning. Quality varies among varieties of fruits and vegetables. Many county Extension offces can recommend varieties best suited for canning. Examine food carefully for freshness and wholesomeness. Discard diseased and moldy food. Trim small diseased lesions or spots from food.

Can fruits and vegetables picked from your garden or purchased from nearby producers when the products are at their peak of quality-within 6 to 12 hours after harvest for most vegetables. For best quality, apricots, nectarines, peaches, pears, and plums should be ripened 1 or more days between harvest and canning. If you must delay the canning of other fresh produce, keep it in a shady, cool place. Fresh home-slaughtered red meats and poultry should be chilled and canned without delay. Do not can meat from sickly or diseased animals. Ice fsh and seafoods after harvest, eviscerate immediately, and can them within 2 days.

Maintaining Color and Favor in Canned Food

To maintain good natural color and favor in stored canned food, you must:

- Remove oxygen from food tissues and jars,
- Quickly destroy the food enzymes,
- Obtain high jar vacuums and airtight jar seals.

Follow these guidelines to ensure that your canned foods retain optimum colors and favors during processing and storage:

- Use only high-quality foods which are at the proper maturity and are free of diseases and bruises.
- Use the hot-pack method, especially with acid foods to be processed in boiling water.
- Don't unnecessarily expose prepared foods to air. Can them as soon as possible.
- While preparing a canner load of jars, keep peeled, halved, quartered, sliced, or diced apples, apricots, nectarines, peaches, and pears in a solution of 3 grams (3,000 milligrams) ascorbic acid to 1 gallon of cold water. This procedure is also useful in maintaining the natural color of mushrooms and potatoes, and for preventing stem-end discoloration in cherries and grapes. You can get ascorbic acid in several forms:

— *Pure powdered form*—seasonally available among canners' supplies in supermarkets. One level teaspoon of pure powder weighs about 3 grams. Use 1 teaspoon per gallon of water as a treatment solution.

— *Vitamin C tablets*—economical and available year-round in many stores. Buy 500-milligram tablets; crush and dissolve six tablets per gallon of water as a treatment solution.

Commercially prepared mixes of ascorbic and citric acid—seasonally available among canners' supplies in supermarkets. Sometimes citric acid powder is sold in supermarkets, but it is less effective in controlling discoloration. If you choose to use these products, follow the manufacturer's directions.

- Fill hot foods into jars and adjust headspace as specifed in recipes.
- Tighten screw bands securely, but if you are especially strong, not as tightly as possible.
- Process and cool jars.
- Store the jars in a relatively cool, dark place, preferably between 50° and 70°F.
- Can no more food than you will use within a year.

Advantages of Hot-packing

Many fresh foods contain from 10 percent to more than 30 percent air. How long canned food retains high quality depends on how much air is removed from food before jars are sealed.

Raw-packing is the practice of flling jars tightly with freshly prepared, but unheated food. Such foods, especially fruit, will foat in the jars. The entrapped air in and around the food may cause discoloration within 2 to 3 months of storage. Raw-packing is more suitable for vegetables processed in a pressure canner.

Hot-packing is the practice of heating freshly prepared food to boiling, simmering it 2 to 5 minutes, and promptly flling jars loosely with the boiled food. Whether food has been hot-packed or raw-packed, the juice, syrup, or water to be added to the foods should also be heated to boiling before adding it to the jars. This practice helps to remove air from food tissues, shrinks food, helps keep the food from foating in the jars, increases vacuum in sealed jars, and improves shelf life. Preshrinking food permits flling more food into each jar.

Hot-packing is the best way to remove air and is the preferred pack style for foods processed in a boiling-water canner. At frst, the color of hot-packed foods may appear no better than that of raw-packed foods, but within a short storage period, both color and favor of hot-packed foods will be superior.

Controlling Headspace

The unflled space above the food in a jar and below its lid is termed headspace. Directions for canning specify leaving 1/4-inch for jams and jellies, 1/2-inch for fruits and tomatoes to be processed in boiling water, and from 1- to 1-1/4-inches in low acid foods to be processed in a pressure canner. This space is needed for expansion of food as jars are processed, and for forming vacuums in cooled jars. The extent of expansion is determined by the air content in the food and by the processing temperature. Air expands greatly when heated to high temperatures; the higher the temperature, the greater the expansion. Foods expand less than air when heated.

Jars and Lids

Food may be canned in glass jars or metal containers. Metal containers can be used only once. They require special sealing equipment and are much more costly than jars.

Regular and wide-mouth Mason-type, threaded, home-canning jars with self-sealing lids are the best choice. They are available in 1/2 pint, pint, 1-1/2 pint, quart, and 1/2 gallon sizes. The standard jar mouth opening is about 2-3/8 inches. Wide-mouth jars have openings of about 3 inches, making them more easily flled and emptied. Half-gallon jars may be used for canning very acid juices. Regular-mouth decorator jelly jars are available in 8 and 12 ounce sizes. With careful use and handling, Mason jars may be reused many times, requiring only new lids each time. When jars and lids are used properly, jar seals and vacuums are excellent and jar breakage is rare.

Most commercial pint- and quart-size mayonnaise or salad dressing jars may be used with new two-piece lids for canning acid foods. However, you should expect more seal failures and jar breakage. These jars have a narrower sealing surface and are tempered less than Mason jars, and may be weakened by repeated contact with metal spoons or knives used in dispensing mayonnaise or salad dressing. Seemingly insignifcant scratches in glass may cause cracking and breakage while processing jars in a canner. Mayonnaise-type jars are not recommended for use with foods to be processed in a pressure canner because of excessive jar breakage. Other commercial jars with mouths that cannot be sealed with two-piece canning lids are not recommended for use in canning any food at home.

Jar Cleaning and Preparation

Before every use, wash empty jars in hot water with detergent and rinse well by hand, or wash in a dishwasher. Unrinsed detergent residues may cause unnatural favors and colors. Jars should be kept hot until ready to fll with food. Submerge the clean empty jars in enough water to cover them in a large stockpot or boiling water canner. Bring the water to a simmer (180°F) and keep the jars in the simmering water until it is time to fll them with food. A dishwasher may be used for preheating jars if they are washed and dried on a complete regular cycle. Keep the jars in the closed dishwasher until needed for flling.

These washing and preheating methods do not sterilize jars. Some used jars may have a white flm on the exterior surface caused by mineral deposits. This scale or hard-water flm on jars is easily removed by soaking jars several hours in a solution containing 1 cup of vinegar (5 percent acidity) per gallon of water prior to washing and preheating the jars.

Sterilization of Empty Jars

All jams, jellies, and pickled products processed less than 10 minutes should be flled into sterile empty jars. To sterilize empty jars after washing in detergent and rinsing thoroughly, submerge them, right side up, in a boiling-water canner with the rack in the bottom. Fill the canner with enough warm water so it is 1 inch above the tops of the jars. Bring the water to a boil, and boil 10 minutes at altitudes of less than 1,000 ft. At higher elevations, boil 1 additional minute for each additional 1,000 ft elevation. Reduce the heat under the canner, and keep the jars in the hot water until it is time to fll them. Remove and drain hot sterilized jars one at a time, saving the hot water in the canner for processing flled jars. Fill the sterilized jars with food, add lids, and tighten screw bands.

Empty jars used for vegetables, meats, and fruits to be processed in a pressure canner need not be presterilized. It is also unnecessary to presterilize jars for fruits, tomatoes, and pickled or fermented foods that will be processed 10 minutes or longer in a boiling-water canner.

Lid selection, preparation, and use

The common self-sealing lid consists of a fat metal lid held in place by a metal screw band during processing. The fat lid is crimped around its bottom edge to form a trough, which is flled with a colored gasket compound. When jars are processed, the lid gasket softens and fows slightly to cover the jar-sealing surface, yet allows air to escape from the jar. The gasket then forms an airtight seal as the jar cools. Gaskets in unused lids work well for at least 5 years from date of manufacture. The gasket compound in older unused lids may fail to seal on jars.

Buy only the quantity of lids you will use in a year. To ensure a good seal, carefully follow the manufacturer's directions in preparing lids for use. Examine all metal lids carefully. Do not use old, dented, or deformed lids, or lids with gaps or other defects in the sealing gasket.

When directions say to fll jars and adjust lids, use the following procedures: After flling jars with food and adding covering liquid, release air bubbles by inserting a fat plastic (not metal) spatula between the food and the jar. Slowly turn the jar and move the spatula up and down to allow air bubbles to escape. (It is not necessary to release air bubbles when flling jams, jellies or all liquid foods such as juices.) Adjust the headspace and then clean the jar rim (sealing surface) with a dampened paper towel. Place the preheated lid, gasket down, onto the cleaned jar-sealing surface. Uncleaned jar-sealing surfaces may cause seal failures. Then ft the metal screw band over the fat lid. Follow the manufacturer's guidelines enclosed with or on the box for tightening the jar lids properly.

Do not retighten lids after processing jars. As jars cool, the contents in the jar contract, pulling the self-sealing lid frmly against the jar to form a high vacuum.

- If rings are too loose, liquid may escape from jars during processing, and seals may fail.
- If rings are too tight, air cannot vent during processing, and food will discolor during storage. Over tightening also may cause lids to buckle and jars to break, especially with raw-packed, pressure-processed food.

Screw bands are not needed on stored jars. They can be removed easily after jars are cooled. When removed, washed, dried, and stored in a dry area, screw bands may be used many times. If left on stored jars, they become diffcult to remove, often rust, and may not work properly again.

Recommended Canners

Equipment for heat-processing home-canned food is of two main types—boiling water canners and pressure canners. Most are designed to hold seven quart jars or eight to nine pints. Small pressure canners hold four-quart jars; some large pressure canners hold 18 pint jars in two layers, but hold only seven quart jars. Pressure saucepans with smaller volume capacities are not recommended for use in canning. Small capacity pressure canners are treated in a similar manner as standard larger canners, and should be vented using the typical venting procedures.

Low-acid foods must be processed in a pressure canner to be free of botulism risks. Although pressure canners may also be used for processing acid foods, boiling water canners are recommended for this purpose because they are faster. A pressure canner would require from 55 to 100 minutes to process a load of jars; while the total time for processing most acid foods in boiling water varies from 25 to 60 minutes. A boiling-water canner loaded with flled jars requires about 20 to 30 minutes of heating before its water begins to boil. A loaded pressure canner requires about 12 to 15 minutes of heating before it begins to vent; another 10 minutes to vent the canner; another 5 minutes to pressurize the canner; another 8 to 10 minutes to process the acid food; and, fnally, another 20 to 60 minutes to cool the canner before remov-ing jars.

Boiling-water Canners

These canners are made of aluminum or porcelain-covered steel. They have removable perforated racks and ftted lids. The canner must be deep enough so that at least 1 inch of briskly boiling water will be over the tops of jars during processing. Some boiling-water canners do not have fat bottoms. A fat bottom must be used on an electric range. Either a fat or ridged bottom can be used on a gas burner. To ensure uniform processing of all jars with an electric range, the canner should be no more than 4 inches wider in diameter than the element on which it is heated.

Using boiling-water Canners

Follow these steps for successful boiling-water canning:

1. Before you start preparing your food, fll the canner halfway with clean water. This is approximately the level needed for a canner load of pint jars. For other sizes and numbers of jars, the amount of water in the canner will need to be adjusted so it will be 1 to 2 inches over the top of the flled jars.
2. Preheat water to 140°F for raw-packed foods and to 180°F for hot-packed foods. Food preparation can begin while this water is preheating.
3. Load flled jars, ftted with lids, into the canner rack and use the handles to lower the rack into the water; or fll the canner with the rack in the bottom, one jar at a time, using a jar lifter. When using a jar lifter, make sure it is securely positioned below the neck of the jar (below the screw band of the lid). Keep the jar upright at all times. Tilting the jar could cause food to spill into the sealing area of the lid.
4. Add more boiling water, if needed, so the water level is at least 1 inch above jar tops. For process times over 30 minutes, the water level should be at least 2 inches above the tops of the jars.
5. Turn heat to its highest position, cover the canner with its lid, and heat until the water in the canner boils vigorously.
6. Set a timer for the total minutes required for processing the food.
7. Keep the canner covered and maintain a boil throughout the process schedule. The heat setting may be lowered a little as long as a complete boil is maintained for the entire process time. If the water stops boiling at any time during the process, bring the water back to a vigorous boil and begin the timing of the process over, from the beginning.
8. Add more boiling water, if needed, to keep the water level above the jars.
9. When jars have been boiled for the recommended time, turn off the heat and remove the canner lid. Wait 5 minutes before removing jars.
10. Using a jar lifter, remove the jars and place them on a towel, leaving at least 1-inch spaces between the jars during cooling. Let jars sit undisturbed to cool at room temperature for 12 to 24 hours.

Pressure Canners

Pressure canners for use in the home have been extensively redesigned in recent years. Models made before the 1970's were heavy-walled kettles with clamp-on or turn-on lids. They were ftted with a dial gauge, a vent port in the form of a petcock or counterweight, and a safety fuse. Modern pressure canners are lightweight, thin walled kettles; most have turn-on lids. They have a jar rack, gasket, dial or weighted gauge, an automatic vent/cover lock, a vent port (steam vent) to be closed with a counterweight or weighted gauge, and a safety fuse.

Pressure does not destroy microorganisms, but high temperatures applied for an adequate period of time do kill microorganisms. The success of destroying all microorganisms capable of growing in canned food is based on the temperature obtained in pure steam, free of air, at sea level. At sea level, a canner operated at a gauge pressure of 10.5 lbs provides an internal temperature of 240°F.

Two serious errors in temperatures obtained in pressure canners occur because:

1. Internal canner temperatures are lower at higher altitudes. To correct this error, canners must be operated at the increased pressures specifed in this publication for appropriate altitude ranges.
2. Air trapped in a canner lowers the temperature obtained at 5, 10, or 15 pounds of pressure and results in under processing. The highest volume of air trapped in a canner occurs in processing raw-packed foods in dial-gauge canners. These canners do not vent air during processing. To be safe, all types of pressure canners must be vented 10 minutes before they are pressurized.

To vent a canner, leave the vent port uncovered on newer models or manually open petcocks on some older models. Heating the flled canner with its lid locked into place boils water and generates steam that escapes through the petcock or vent port. When steam frst escapes, set a timer for 10 minutes. After venting 10 minutes, close the petcock or place the counterweight or weighted gauge over the vent port to pressurize the canner.

Weighted-gauge models exhaust tiny amounts of air and steam each time their gauge rocks or jiggles during processing. They control pressure precisely and need neither watching during processing nor checking for accuracy. The sound of the weight rocking or jiggling indicates that the canner is maintaining the recommended pressure. The single disadvantage of weighted-gauge canners is that they cannot correct precisely for higher altitudes. At altitudes above 1,000 feet, they must be operated at canner pressures of 10 instead of 5, or 15 instead of 10, PSI.

Check dial gauges for accuracy before use each year. Gauges that read high cause under-processing and may result in unsafe food. Low readings cause over-processing. Pressure adjustments can be made if the gauge reads up to 2 pounds high or low. Replace gauges that differ by more than 2 pounds. Every pound of pressure is very important to the temperature needed inside the canner for producing safe food, so accurate gauges and adjustments are essential when a gauge reads higher than it should. If a gauge is reading lower than it should, adjustments may be made to avoid overprocessing, but are not essential to safety. Gauges may be checked at many county Cooperative Extension offces or contact the pressure canner manufacturer for other options.

Handle canner lid gaskets carefully and clean them according to the manufacturer's directions. Nicked or dried gaskets will allow steam leaks during pressurization of canners. Keep gaskets clean between uses. Gaskets on older model canners may require a light coat of vegetable oil once per year. Gaskets on newer model canners are pre-lubricated and do not beneft from oiling. Check your canner's instructions if there is doubt that the particular gasket you use has been pre-lubricated.

Lid safety fuses are thin metal inserts or rubber plugs designed to relieve excessive pressure from the canner. Do not pick at or scratch fuses while cleaning lids. Use only canners that have the Underwriter's Laboratory (UL) approval to ensure their safety.

Replacement gauges and other parts for canners are often available at stores offering canning equipment or from canner manufacturers. When ordering parts, give your canner model number and describe the parts needed.

Using pressure canners

Follow these steps for successful pressure canning:

1. Put 2 to 3 inches of hot water in the canner. Some specifc products in this Guide require that you start with even more water in the canner. Always follow the directions with USDA processes for specifc foods if they require more water added to the canner. Place flled jars on the rack, using a jar lifter. When using a jar lifter, make sure it is securely

positioned below the neck of the jar (below the screw band of the lid). Keep the jar upright at all times. Tilting the jar could cause food to spill into the sealing are of the lid. Fasten canner lid securely.

2. Leave weight off vent port or open petcock. Heat at the highest setting until steam fows freely from the open petcock or vent port.
3. While maintaining the high heat setting, let the steam fow (exhaust) continuously for 10 minutes, and then place the weight on the vent port or close the petcock. The canner will pressurize during the next 3 to 5 minutes.
4. Start timing the process when the pressure reading on the dial gauge indicates that the recommended pressure has been reached, or when the weighted gauge begins to jiggle or rock as the canner manufacturer describes.
5. Regulate heat under the canner to maintain a steady pressure at or slightly above the correct gauge pressure. Quick and large pressure variations during processing may cause unnecessary liquid losses from jars. Follow the canner manufacturer's directions for how a weighted gauge should indicate it is maintaining the desired pressure.
6. When the timed process is completed, turn off the heat, remove the canner from heat if possible, and let the canner depressurize. Do not force-cool the canner. Forced cooling may result in unsafe food or food spoilage. Cooling the canner with cold running water or opening the vent port before the canner is fully depressurized will cause loss of liquid from jars and seal failures. Force-cooling may also warp the canner lid of older model canners, causing steam leaks. Depressurization of older models without dial gauges should be timed. Standard-size heavy-walled canners require about 30 minutes when loaded with pints and 45 minutes with quarts. Newer thin-walled canners cool more rapidly and are equipped with vent locks. These canners are depressurized when their vent lock piston drops to a normal position.
7. After the canner is depressurized, remove the weight from the vent port or open the petcock. Wait 10 minutes, unfasten the lid, and remove it carefully. Lift the lid away from you so that the steam does not burn your face.

8. Remove jars with a jar lifter, and place them on a towel, leaving at least 1-inch spaces between the jars during cooling. Let jars sit undisturbed to cool at room temperature for 12 to 24 hours.

Selecting the correct processing time

When canning in boiling water, more processing time is needed for most raw-packed foods and for quart jars than is needed for hot-packed foods and pint jars.

To destroy microorganisms in acid foods processed in a boiling-water canner, you must:

- Process jars for the correct number of minutes in boiling water.
- Cool the jars at room temperature.

The food may spoil if you fail to add process time for lower boiling-water temperatures at altitudes above 1,000 feet, process for fewer minutes than specifed, or cool jars in cold water.

To destroy microorganisms in low-acid foods processed with a pressure canner, you must:

- Process the jars using the correct time and pressure specifed for your altitude.
- Allow canner to cool at room temperature until it is completely depressurized.

The food may spoil if you fail to select the proper process times for specifc altitudes, fail to exhaust canners properly process at lower pressure than specifed, process for fewer minutes than specifed, or cool the canner with water.

Using tables for determining proper process times

This set of guides includes processing times with altitude adjustments for each product. Process times for 1/2-pint and pint jars are the same, as are times for 1-1/2 pint and quart jars. For some products, you have a choice of processing at 5, 10, or 15 PSI. In these cases, choose the canner pressure you wish to use and match it with your pack style (raw or hot) and jar size to fnd the correct process time.

Cooling Jars

When you remove hot jars from a canner, do not retighten their jar lids.

Retightening of hot lids may cut through the gasket and cause seal failures. Cool the jars at room temperature for 12 to 24 hours. Jars may be cooled on racks or towels to minimize heat damage to counters. The food level and liquid volume of raw-packed jars will be noticeably lower after cooling. Air is exhausted during processing and food shrinks. If a jar loses excessive liquid during processing, do not open it to add more liquid. Check for sealed lids as described below.

Testing jar seals

After cooling jars for 12 to 24 hours, remove the screw bands and test seals with one of the following options:

— *Option 1*. Press the middle of the lid with a fnger or thumb. If the lid springs up when you release your fnger, the lid is unsealed.

— *Option 2*. Tap the lid with the bottom of a teaspoon. If it makes a dull sound, the lid is not sealed. If food is in contact with the underside of the lid, it will also cause a dull sound. If the jar is sealed correctly, it will make a ringing, high-pitched sound.

— *Option 3*. Hold the jar at eye level and look across the lid. The lid should be concave (curved down slightly in the center). If center of the lid is either fat or bulging, it may not be sealed.

Reprocessing Unsealed Jars

If a lid fails to seal on a jar, remove the lid and check the jar-sealing surface for tiny nicks. If necessary, change the jar, add a new, properly prepared lid, and reprocess within 24 hours using the same processing time. Headspace in unsealed jars may be adjusted to 1-1/2 inches and jars could be frozen instead of reprocessed. Foods in single unsealed jars could be stored in the refrigerator and consumed within several days.

Storing Canned Foods

If lids are tightly vacuum sealed on cooled jars, remove screw bands, wash the lid and jar to remove food residue; then rinse and dry jars. Label and date the jars and store them in a clean, cool, dark, dry place. Do not store jars above 95°F or near hot pipes, a range, a furnace, under a sink, in an uninsulated attic, or in direct sunlight. Under these conditions, food will lose quality in a few weeks or months and may spoil. Dampness may corrode metal lids, break seals, and allow recontamination and spoilage.

Accidental freezing of canned foods will not cause spoilage unless jars become unsealed and recontaminated. However, freezing and thawing may soften food. If jars must be stored where they may freeze, wrap them in newspapers, place them in heavy cartons, and cover with more newspapers and blankets.

Identifying and Handling Spoiled Canned Food

Do not taste food from a jar with an unsealed lid or food that shows signs of spoilage. You can more easily detect some types of spoilage in jars stored without screw bands. Growth of spoilage bacteria and yeast produces gas which pressurizes the food, swells lids, and breaks jar seals. As each stored jar is selected for use, examine its lid for tightness and vacuum. Lids with concave centers have good seals.

Next, while holding the jar upright at eye level, rotate the jar and examine its outside surface for streaks of dried food originating at the top of the jar. Look at the contents for rising air bubbles and unnatural color.

While opening the jar, smell for unnatural odors and look for spurting liquid and cotton-like mold growth (white, blue, black, or green) on the top food surface and underside of lid.

Spoiled low-acid foods, including tomatoes, may exhibit different kinds of spoilage evidence or very little evidence. Therefore, all suspect containers of spoiled low-acid foods, including tomatoes, should be treated as having produced botulinum toxin and handled carefully in one of two ways:

- If the suspect glass jars or swollen metal cans are still sealed, place them in a heavy garbage bag. Close and place the bag in a regular trash container or dispose in a nearby landfll.
- If the suspect glass jars or cans are unsealed, open, or leaking, they should be detoxifed before disposal.

Detoxifcation process: Wear disposable rubber or heavy plastic gloves. Carefully place the suspect containers and lids on their sides in an 8-quart volume or larger stock pot, pan, or boiling-water canner. Wash your hands with gloves thoroughly. Carefully add water to the pot and avoid splashing the water. The water should completely cover the containers with a minimum of a 1-inch level above the containers. Place a lid on the pot and heat the water to boiling. Boil 30 minutes to ensure detoxifying the food and all container components. Cool and discard the containers, their lids, and food in the trash or dispose in a nearby landfll.

Cleaning up the area: Contact with botulinum toxin can be fatal whether it is ingested or enters through the skin. Take care to avoid contact with suspect foods or liquids. Wear rubber or heavy plastic gloves when handling suspect foods or cleaning up contaminated work surfaces and equipment. A fresh solution of 1 part unscented liquid household chlorine bleach (5 to 6% sodium hypochlorite) to 5 parts clean water should be used to treat work surfaces, equipment, or other items, including can openers and clothing, that may have come in contact with suspect foods or liquids. Spray or wet contaminated surfaces with the bleach solution and let stand for 30 minutes. Wearing gloves, wipe up treated spills with paper towels being careful to minimize the spread of contamination. Dispose of these paper towels by placing them in a plastic bag before putting them in the trash. Next, apply the bleach solution to all surfaces and equipment again, and let stand for 30 minutes and rinse. As a last step, thoroughly wash all detoxifed counters, containers, equipment, clothing, etc. Discard gloves when cleaning process is complete.

Preparing Pickled and Fermented Foods

The many varieties of pickled and fermented foods are classifed by ingredients and method of preparation.

Regular dill pickles and sauerkraut are fermented and cured for about 3 weeks. Refrigerator dills are fermented for about 1 week. During curing, colors and favors change and acidity increases. Fresh-pack or quick-process pickles are not fermented; some are brined several hours or overnight, then drained and covered with vinegar and seasonings. Fruit pickles usually are prepared by heating fruit in a seasoned syrup acidifed with either lemon juice or vinegar. Relishes are made from chopped fruits and vegetables that are cooked with seasonings and vinegar.

Be sure to remove and discard a 1/16-inch slice from the blossom end of fresh cucumbers. Blossoms may contain an enzyme which causes excessive softening of pickles.

Caution: The level of acidity in a pickled product is as important to its safety as it is to taste and texture.

- Do not alter vinegar, food, or water proportions in a recipe or use a vinegar with unknown acidity.
- Use only recipes with tested proportions of ingredients.

- There must be a minimum, uniform level of acid throughout the mixed product to prevent the growth of botulinum bacteria.

  Ingredients

Select fresh, frm fruits or vegetables free of spoilage. Measure or weigh amounts carefully, because the proportion of fresh food to other ingredients will affect favor and, in many instances, safety.

Use canning or pickling salt. Noncaking material added to other salts may make the brine cloudy. Since fake salt varies in density, it is not recommended for making pickled and fermented foods. White granulated and brown sugars are most often used. Corn syrup and honey, unless called for in reliable recipes, may produce undesirable favors. White distilled and cider vinegars of 5 percent acidity (50 grain) are recommended. White vinegar is usually preferred when light color is desirable, as is the case with fruits and caulifower.

Pickles with reduced salt content

In the making of fresh-pack pickles, cucumbers are acidifed quickly with vinegar. Use only tested recipes formulated to produce the proper acidity. While these pickles may be prepared safely with reduced or no salt, their quality may be noticeably lower. Both texture and favor may be slightly, but noticeably, different than expected. You may wish to make small quantities frst to determine if you like them.

However, the salt used in making fermented sauerkraut and brined pickles not only provides characteristic favor but also is vital to safety and texture. In fermented foods, salt favors the growth of desirable bacteria while inhibiting the growth of others. Caution: Do not attempt to make sauerkraut or fermented pickles by cutting back on the salt required.

Firming agents

Alum may be safely used to frm fermented pickles. However, it is unnecessary and is not included in the recipes in this publication. Alum does not improve the frmness of quick-process pickles. The calcium in lime defnitely improves pickle frmness. Food-grade lime may be used as a lime-water solution for soaking fresh cucumbers 12 to 24 hours before pickling them. Excess lime absorbed by the cucumbers must be removed to make safe pickles. To remove excess lime, drain the lime-water solution, rinse, and then resoak the cucumbers in fresh water for 1 hour. Repeat the rinsing and soaking steps two more times. To further improve pickle firmness, you may process cucumber pickles for 30 minutes in water at 180°F. This process also prevents spoilage, but the water temperature should not fall below 180°F. Use a candy or jelly thermometer to check the water temperature.

Preventing spoilage

Pickle products are subject to spoilage from microorganisms, particularly yeasts and molds, as well as enzymes that may affect favor, color, and texture. Processing the pickles in a boiling-water canner will prevent both of these problems. Standard canning jars and self-sealing lids are recommended. Processing times and procedures will vary according to food acidity and the size of food pieces.

Preparing Butters, Jams, Jellies, and Marmalades

Sweet spreads are a class of foods with many textures, favors, and colors. They all consist of fruits preserved mostly by means of sugar and they are thickened or jellied to some extent. Fruit jelly is a semi-solid mixture of fruit juice and sugar that is clear and frm enough to hold its shape. Other spreads are made from crushed or ground fruit.

Jam also will hold its shape, but it is less frm than jelly. Jam is made from crushed or chopped fruits and sugar. Jams made from a mixture of fruits are usually called conserves, especially when they include citrus fruits, nuts, raisins, or coconut. Preserves are made of small, whole fruits or uniform-size pieces of fruits in a clear, thick, slightly jellied syrup. Marmalades are soft fruit jellies with small pieces of fruit or citrus peel evenly suspended in a transparent jelly. Fruit butters are made from fruit pulp cooked with sugar until thickened to a spreadable consistency.

Ingredients

For proper texture, jellied fruit products require the correct combination of fruit, pectin, acid, and sugar. The fruit gives each spread its unique favor and color. It also supplies the water to dissolve the rest of the necessary ingredients and furnishes some or all of the pectin and acid. Good-quality, favorful fruits make the best jellied products.

Pectins are substances in fruits that form a gel if they are in the right combination with acid and sugar. All fruits contain some pectin. Apples, crab apples, gooseberries, and some plums and grapes usually contain enough natural pectin to form a gel. Other fruits, such as strawberries, cherries, and blueberries, contain little pectin and must be combined with other fruits high in pectin or with commercial pectin products to obtain gels. Because fully ripened fruit has less pectin, one-fourth of the fruit used in making jellies without added pectin should be underripe.

Caution: Commercially frozen and canned juices may be low in natural pectins and make soft textured spreads.

The proper level of acidity is critical to gel formation. If there is too little acid, the gel will never set; if there is too much acid, the gel will lose liquid (weep). For fruits low in acid, add lemon juice or other acid ingredients as directed. Commercial pectin products contain acids which help to ensure gelling.

Sugar serves as a preserving agent, contributes favor, and aids in gelling. Cane and beet sugar are the usual sources of sugar for jelly or jam. Corn syrup and honey may be used to replace part of the sugar in recipes, but too much will mask the fruit favor and alter the gel structure. Use tested recipes for replacing sugar with honey and corn syrup. Do not try to reduce the amount of sugar in traditional recipes. Too little sugar prevents gelling and may allow yeasts and molds to grow.

Jams and jellies with reduced sugar

Jellies and jams that contain modifed pectin, gelatin, or gums may be made with noncaloric sweeteners. Jams with less sugar than usual also may be made with concentrated fruit pulp, which contains less liquid and less sugar.

Two types of modifed pectin are available for home use. One gels with one-third less sugar. The other is a low-methoxyl pectin which requires a source of calcium for gelling. To prevent spoilage, jars of these products may need to be processed longer in a boiling-water canner. Recipes and processing times provided with each modifed pectin product must be followed carefully. The proportions of acids and fruits should not be altered, as spoilage may result. Acceptably gelled refrigerator fruit spreads also may be made with gelatin and sugar substitutes. Such products spoil at room temperature, must be refrigerated, and should be eaten within 1 month.

Preventing spoilage

Even though sugar helps preserve jellies and jams, molds can grow on the surface of these products. Research now indicates that the mold which people usually scrape off the surface of jellies may not be as harmless as it seems. Mycotoxins have been found in some jars of jelly having surface mold growth. Mycotoxins are known to cause cancer in animals; their effects on humans are still being researched. Because of possible mold contamination, paraffn or wax seals are no longer recommended for any sweet spread, including jellies. To prevent growth of molds and loss of good favor or color, fll products hot into sterile Mason jars, leaving 1/4-inch headspace, seal with self-sealing lids, and process 5 minutes in a boiling-water canner. Correct process time at higher elevations by adding 1 additional minute per 1,000 ft above sea level. If unsterile jars are used, the flled jars should be processed 10 minutes. Use of sterile jars is preferred, especially when fruits are low in pectin, since the added 5-minute process time may cause weak gels.

Methods of Making Jams and Jellies

The standard method, which does not require added pectin, works best with fruits naturally high in pectin. The other method, which requires the use of commercial liquid or powdered pectin, is much quicker. The gelling ability of various pectins differs. To make uniformly gelled products, be sure to add the quantities of commercial pectins to specifc fruits as instructed on each package. Overcooking may break down pectin and prevent proper gelling. When using either method, make one batch at a time, according to the recipe. Increasing the quantities often results in soft gels. Stir constantly while cooking to prevent burning. Recipes are developed for specifc jar sizes. If jellies are flled into larger jars, excessively soft products may result.

Canned Foods for Special Diets

The cost of commercially canned special diet food often prompts interest in preparing these products at home. Some low-sugar and low-salt foods may be easily and safely canned at home. However, the color, flavor, and texture of these foods may be different than expected and be less acceptable

Canning without sugar

In canning regular fruits without sugar, it is very important to select fully ripe but firm fruits of the best quality. Prepare these as described for hot-packs in Guide 2, but use water or regu-lar unsweetened fruit juices instead of sugar syrup. Juice made from the fruit being canned is best. Blends of unsweetened apple, pineapple, and white grape juice are also good for filling over solid fruit pieces. Adjust headspaces and lids and use the processing recommendations given for regular fruits. Splenda® is the only sugar substitute currently in the marketplace that can be added to covering liquids before canning fruits. Other sugar substitutes, if desired, should be added when serving.

Canning without salt (reduced sodium)

To can tomatoes, vegetables, meats, poultry, and seafood, use the procedures given in Guides 3 through 5, but omit the salt. In these products, salt seasons the food but is not necessary to ensure its safety. Add salt substitutes, if desired, when serving.

Canning Fruit-based Baby Foods

You may prepare any chunk-style or pureed fruit with or without sugar, using the procedure for preparing each fruit. Pack in half-pint, preferably, or pint jars and use the following processing times.

Caution: Do not attempt to can pureed vegetables, red meats, or poultry meats, because proper processing times for pureed foods have not been determined for home use.

Instead, can and store these foods using the standard processing procedures; puree or blend them at serving time. Heat the blended foods to boiling, simmer for 10 minutes, cool, and serve. Store unused portions in the refrigerator and use within 2 days for best quality.

References

Great Britain, Department of Health and Social Security, 1981. *The Canning of Low-acid Foods: A Guide to Food Manufacturing Practice.* London.

Potter, N. N. and Hotchkiss, J. H. 1999. *Food Science.* Springer.

Fellows, P. J. 1999. *Food Processing Technology: Principles and Practice.* Woodhead Pub.

Stumbo, C.R., 1973. *Thermobacteriology in food processing.* New York, Academic Press.

Bibliography

Aitken,A *et al.* (eds) 1982. *Fish Handling and Processing*. Second Edition, Edinburgh, Her Majesty's Stationery Office, £10.

Alasalvar C. 2010. *Seafood Quality, Safety and Health Applications* John Wiley and Sons.

Alzamora SM, Tapia MS and López-Malo A. 2000. *Minimally processed fruits and vegetables: fundamental aspects and applications* Springer.

Arthey, D. and Ashurst, P.R. 1996. *Fruit Processing*. Blackie Academic & Professional, London, 142-125.

Banwart, G.J. 1979. *Basic Food Microbiology*, AVI Publishing Co. Inc., Westport, Connecticut.

Bender, AK. and Zia, M. 1976. *Meat quality and protein quality*. 1. Fd. Technol. 11. 495498.

FAO. 1990. *Manual of simple methods of meat preservation. F*AO Animal Production and Health Paper No. 79. Rome, FAO.

Fellows, P. J. 1999. *Food Processing Technology: Principles and Practice*. Woodhead Pub.

Fennema, O. 1975. *Effect of freeze-preservation of food processing*. pp 244-288. Ed. R.S. Harris and Karmas, E. Avi Publishing Co. Westport, Conn.

Frazier, W.C. and Westoff, D.C. 1996. *Food Microbiology,* Tata McGraw Hill Publishing Co. Ltd., New Delhi.

Garbutt, J. 1998. *Essentials of Food Microbiology*, Arnold International Student's Edition, London.

Great Britain, Department of Health and Social Security, 1981. *The Canning of Low-acid Foods: A Guide to Food Manufacturing Practice*. London.

Henkel, J. 1998. Irradiation – A safe measure for safer foods. *FDA Consumer,* May–June, Pages 12-17.

Hudson, R.J. 1976. Potential for Meat Production from Marginal Land Resources. *Misc. Publ. Dept. of Animal Science.* University of Alberta, Edmonton, Alberta, Canada

ICGFI. 1998 "Irradiation and Trade in Food and Agriculture Products". *International Consultative Group on Food Irradiation Policy Document,* Vienna.

International Institute of Refrigeration. *Guide to Refrigerated Storage*. Guide de Paris,

Islami, F. 2011. "Pickled vegetables and the risk of oesophageal cancer: a meta-analysis". *British Journal of Cancer*. Retrieved. 2011-08-16.

Jenkins, C.H. 1968. *Modern Warehouse Management.* New York, McGraw-Hill Book Company

Karmas, E. 1976. *Processed meat technology*. New Jersey, USA, Noyes Data Corporation.

Lee S. 2004. "Microbial Safety of Pickled Fruits and Vegetables and Hurdle Technology" *Internet Journal of Food Safety*, 4: 21-32.

Leistner I. 2000. "Basic aspects of food preservation by hurdle technology" *International Journal of Food Microbiology*, 55:181–186.

McGee, Harold. 2004. *On Food and Cooking: The Science and Lore of the Kitchen*. New York: Scribner, pp. 291–296.

Myers, M., 1981, Planning and Engineering Data. 2. *Fresh Fish Handling*. FAO Fish. Circ., (735), 64p.

Nagy, S. and Shaw, P.E.. *Tropical and Subtropical Fruits. Composition, Properties and uses*. AVI Publishing, Westport, Connecticut, 127.

Pelczar, M. Jr., Chan, E.C.S. and Kreig, N.R. 1993. *Microbiology*, Tata McGraw Hill Inc., New York.

Potter, N. N. and Hotchkiss, J. H. 1999. *Food Science*. Springer.

Ryall, A.L. and Pentzer, W.T. 1979. *Handling, Transportation and Storage of Fruits and Vegetables*, 320-332.

Satin, M. 1993. *Food Irradiation–A Guidebook*. Technomic Publishing Co., Inc., Lancaster, USA.

Sofos, J.N. 1989. *Sorbate Food Preservatives*. CRC Press, Boca Raton, 16-17.

Stumbo, C.R., 1973. *Thermobacteriology in food processing*. New York, Academic Press.

Thakur, B.R. and Singh, R.K. 1994. Food irradiation – chemistry and applications. *Food Reviews International*, Volume 10, Part 4, Pages 437-473.

Thompson, A.K. 1996. *Post-harvest Technology of Fruit and Vegetables*. Blackwell Science, Berlin.

WHO. 1988. *Food Irradiation. A Technique for Preserving and Improving the Safety of Food*. World Health Organization, Geneva.

Yousef AE and Carolyn Carlstrom C. 2003. *Food microbiology: a laboratory manual* Wiley.

Zeldes, Leah A. 2009. "Eat this! Southern-fried dill pickles, a rising trend". *Dining Chicago*. Chicago's Restaurant & Entertainment Guide, Inc.. Retrieved 2010-08-02.